Rotating Machines

Edited by Yan Li

Published in London, United Kingdom

Rotating Machines
http://dx.doi.org/10.5772/intechopen.97952
Edited by Yan Li

Contributors
Yuning Zhang, Hao Li, Xianghao Zheng, Jinwei Li, Salha Faleh, Tahar Khir, Minnan Yue, Peilin Wang, Chun Li, Kailun Niu, Yangtian Yan, Xinyu Pei, Yan Li, Zhongqiu Mu, Zhiyuan Liu, Wenfeng Guo, Fang Feng, Kotaro Tagawa, Tieliu Jiang, Shengwen Wang, Zhongbin Zhang, Lidong Zhang

First published in London, United Kingdom, 2022 by IntechOpen
IntechOpen is the global imprint of INTECHOPEN LIMITED, registered in England and Wales, registration number: 11086078, 5 Princes Gate Court, London, SW7 2QJ, United Kingdom

British Library Cataloguing-in-Publication Data
A catalogue record for this book is available from the British Library

Additional hard and PDF copies can be obtained from orders@intechopen.com

Rotating Machines
Edited by Yan Li
p. cm.
Print ISBN 978-1-80355-735-9
Online ISBN 978-1-80355-736-6
eBook (PDF) ISBN 978-1-80355-737-3

Meet the editor

Dr. Yan Li is a professor at the Engineering College, Northeast Agricultural University, China. He has more than fifteen years of research and teaching experience. His research interests include energy engineering, agricultural engineering, and mechanical engineering. He teaches both undergraduate and graduate courses in new energy, fluid dynamics, and wind energy engineering. His current research focus is renewable energy, especially wind energy. He has more than fifty publications on wind turbines, wind turbine icing, and anti-icing technology to his credit. Dr. Li is a member of the Chinese Aerodynamics Research Society, Chinese Society of Engineering Thermophysics, and Chinese Wind Energy Association.

Contents

Preface

Rotating machines are machines that mainly rely on rotating action to complete certain functions of fluid transportation, energy conversion, and power generation. There are many kinds of rotating machines. The typical rotating machines include turbines, compressors, fans, pumps, generators, and motors, which are widely used in electric power, petrochemical, metallurgy, aerospace, and other fields.

The research and application of rotating machines have a long history. In recent decades, energy and environmental issues have become increasingly prominent, and the development and utilization of new energy have been widely valued and developed rapidly. The application of rotating machines in the field of new and renewable energy has become an important development direction. We can often see the beautiful scenery of many wind turbines rotating on the mountains, plains, and sea. This book focuses on the recent research progress and development of rotating machines technology in the energy field. It includes four research reports and one review.

A wind turbine is a device that uses the wind to drive the turbine rotor to rotate to convert the kinetic energy of the wind into mechanical energy, electric energy, and heat energy. Wind energy is one of the fastest-growing and most widely used renewable energies in the world. This book includes three research reports on offshore wind turbines, wind heating by vertical axis wind turbines, and icing impacts on wind turbine blades.

Offshore wind turbines are being increasingly used for generating large-scale wind energy. The application of this technology in the deep sea is an important research field. Factors such as waves, earthquakes, extrema weather, and so on affect the performance of offshore wind turbines. Therefore, their foundation structure is very important. This book compares the structural dynamic response between jacket foundation large-scale offshore wind turbines and monopile ones under wind and seismic loads.

In addition to wind electric power generation, wind heating is also a very important direction of wind energy conversion and utilization. Direct heating of a wind turbine is a form of heating production using wind energy with the advantages of low requirements for wind quality, relatively simple device structure, and high heating efficiency. Vertical axis wind turbines are another important form of a wind turbine. They have characteristics of wind direction independence, novel structure, and easy maintenance, and are suitable for small-scale wind energy utilization. The book includes numerical simulation and wind tunnel tests of heating systems under different working conditions. It also examines the adaptability of the wind turbine and the heating efficiency of the heating device.

Wind turbines sometimes face a variety of extreme weather conditions, such as icing, heavy snow, lightning, sandstorms, and so on. For the wind turbines installed in cold and humid regions, the icing on the blade surface will often occur, which affects the aerodynamic characteristics and safe operation of the turbine rotor and thus the output performance of electric power of wind farms. This book discusses icing and its impacts on the blade surfaces of wind turbines. It also presents the results of wind tunnel testing of icing on a 2D blade airfoil used for wind turbines. The study tested and analyzed the icing distributions on the blade surface at different attack angles under different ambient temperatures and wind speeds.

Pump turbines are yet another type of rotating machine. A reversible pump-turbine is the core part of a pumped hydro energy storage power station, which can switch between pumping and generating modes based on demand. Signal analysis methods have been widely applied to monitor the condition and fault diagnosis of various kinds of rotating machinery, especially pump turbines. This book reviews signal analysis methods and their applications in reversible pump turbines, including time-domain analysis, time-frequency analysis, mode decomposition, and de-noising. According to this review, signal analysis and processing can help us better understand the operational states of reversible pump turbines. Through mode decomposition of the signal, various components within the signal can be distinguished and signal de-noising can be performed.

A gas turbine is the core component of a power plant. To obtain high efficiency, many new kinds of gas turbines have been proposed. The SOFC-GT hybrid system, which is a combination of solid oxide fuel cells and a gas turbine, is one example. This book presents an exergy analysis of the SOFC-TG system. The exergy balances are established for the different components. According to the results, the exergy efficiencies are improved by increasing the pre-reforming fraction of the air and fuel flows.

We can foresee that rotating machines will play a greater role in more fields in the future. The research on rotating machines will be more in-depth and innovative.

Yan Li
Engineering College,
Northeast Agricultural University,
Harbin, China

Chapter 1

Comparative Analysis of Transient Dynamics of Large-Scale Offshore Wind Turbines with Different Foundation Structure under Seismic

Peilin Wang, Minnan Yue, Chun Li, Yangtian Yan, Kailun Niu and Xinyu Pei

Abstract

In this paper, a structural dynamic response comparison between jacket foundation large-scale offshore wind turbines (OWTs) and monopile ones under wind and seismic loads is demonstrated. The interaction between flexible soil and pile foundation is described by Winkler soil-structure interaction (SSI) model. The National Renewable Energy Laboratory (NREL) 5 MW large-scale OWT is studied via the finite element model. The structural transient dynamic response of these two structures under normal operating conditions at rated wind speed and earthquake is calculated. The results show that under the action of seismic and turbulent wind, the jacket has better wind and seismic resistance, and the displacement of the top of the tower is small, which can effectively protect the blades and the nacelle. Compared with monopile, the range of the Mises equivalent stress amplitude of the jacket wind turbine was reduced and the average value was decreased at the time of the seismic. The study also found that the existing jacket design will have a local strain energy surge.

Keywords: OWTs, dynamic response, SSI, jacket foundation, monopile foundation, seismic

1. Introduction

The need for renewable resources is becoming more evident as the earth's limited reserves of remaining fossil fuels have essentially halted development and utilization. Among renewable energy sources such as wind energy, solar energy, tidal energy and geothermal energy, wind energy is considered to be the most cost-effective alternative energy and one of the cheapest new power source, showing its great potential to meet demand [1]. The 2021 Global Wind Energy Council (GWEC) released an annual wind report that 2020 was the best year in history for the global wind industry showing year-over-year growth of 53%. Installing more than 93 GW wind power in a challenging year with disruption to both the global supply chain and project construction has demonstrated the incredible resilience of the wind industry. According to the GWEC, more than 6 GW of new offshore wind power capacity was installed worldwide in 2020, with China adding more than 3 GW of new offshore wind power capacity, accounting for half of the world's new installed capacity [2]. These estimations reasonably take into account the new improved technologies in wind turbine systems and the offshore wind energy available in considerable quantity. Those sites primarily located at near coastal or offshore areas and experiencing high wind throughout the year, such as the sea off southeast China and the western coast of the United States. However, these areas are with close proximity to the Pacific seismic belts, making them vulnerable to severe and frequent earthquakes [3]. Or rather, OWTs working under such condition generally have to experience dual external loadings primarily stimulated by turbulent wind and seismic excitation. In addition, so as to shrink the cost of energy, the size of OWTs has increased to 9.5 MW recently [4]. Nevertheless, higher wind turbine capacity is accompanied by higher wind turbine towers, which are destined to withstand higher wind pressures and are more vulnerable to seismic loads.

In the past few decades, a large number of studies have been carried out to analyze the dynamic behavior of OWTs under earthquake excitation, in order to improve the stability of wind turbines [5–9], but there is still the issue of oversimplification in the geometry of the model. It is noted that in most of the aforementioned literatures, the rotor and nacelle were either completely ignored or simplified as a lumped mass. In Ref. [10–13], Asareh and Prowell selected FAST as a design basis, developed a seismic module to examine the coupled effect of wind and seismic loads, and employed an improved FAST (also known as NREL Seismic) to investigate the relationship between seismic intensity and structural response. Additionally, in Ref. [14–17], the natural frequency and mode shape of the OWT support structure are calculated using frequency domain, and then the seismic load is calculated according to the seismic response spectrum. However, unsteady aerodynamic loads were often ignored or treated as a rotor thrust, resulting in relatively inaccurate prediction of combined loads acting on blades.

In Ref. [18], the coupled behaviors of seismic and wind loads based on a 2 MW wind turbine was conducted by Witcher. In Ref. [19], Jin et al. also employed the NREL Seismic tool to predict the dynamic responses under multiple hazards associated with earthquake excitation and turbulent wind. In general, the aerodynamic load of large-scale wind turbines increases exponentially with the rotor diameter. Hence, the continuous aerodynamic effects on a larger turbine are determined to be unneglectable in the comparative study of operational and parked states. In Ref. [20], the dynamic response of a 5 MW OWT under seismic loading was further discussed. The finite element model of the SSI was considered for the multi-degree of freedom of fixed foundation. Response spectrum analysis method and transient dynamics analysis method were adopted to verify that the SSI cannot be ignored. In addition, some scholars assumed that there is rigid or linear elastic contact between seabed soil and pile foundation [21–25]. However, due to the severe liquefaction of the seabed soil, the soil porosity and water content change continuously with the pile depth. Furthermore, mechanical dynamic responses on the surface exhibit nonlinear changes in both the horizontal and vertical directions of soil, which can raise the damping of the OWT support structure and lessen the natural frequency. In this case, the relative displacement between soil and structure intensifies the liquefaction of soil during earthquake, and the soil reaction shows highly nonlinear characteristics related to the structural deformation difference [26–33]. To sum up, an accurate nonlinear SSI model is needed to analyze the dynamic characteristic and response of the OWT support structures under mixed loads.

To achieve the same end, Yang et al. [34] proposed a numerical analysis framework coupled with FAST to obtain more precise responses. However, this method is only applicable to OWTs with monopile support structure, and the results are relatively macroscopic. Yet, it is challenging to examine the local response characteristics of OWTs with different support structures under joint loads.

To tackle these challenges, it is necessary to propose a proper and practical control mechanism of large-scale OWTs under persistent earthquake-and-wind-induced excitations, so as to decrease the volume, prolong the working life, and improve the reliability of the support structures. In the paper, instead of 2 MW OWT, the NREL 5 MW OWT is referred to as the research object, utilizing Winkler soil-structure interaction (SSI) model to describe the interaction between flexible soil and pile foundation. Finite element analysis software ANSYS is used to simulate the dynamic response of OWT foundation under different soils and different seismic intensities, and the influence of seismic load on tower-top displacement and tower-base bending moment is studied. Moreover, the local response characteristics of OWTs with different support structures under seismic load are strictly analyzed. Several principles of nonlinear dynamic response of OWTs structures experiencing sorts of earthquakes are identified. These conclusions will be beneficial in the phase of guaranteeing the optimum performance of OWTs foundations in any operating conditions and in any seismic area.

The remainder of the paper is presented as follows. In Section I, numerical modeling and physical properties are explained. Section II describes wind and seismic loads. In Section III, the finite element method is outlined. Reliability is validated in Section IV. The comparation between the numerical results of two foundations are presented in Section V. Finally, Section VI summarizes the work done and concludes the paper.

2. Numerical simulations

This paper takes NREL 5 MW OWT as the research object. The support structures of offshore wind power are mainly monopile and jacket, which are usually installed in shallow sea areas. The jacket is mainly composed of grid truss, low steel consumption, convenient transportation and assembly, and good anti-wind and wave performance, suitable for deep water [35]. Since this paper mainly studies the difference between monopile and jacket under the combined action of earthquake and turbulent wind, and wave load has little influence on the jacket, the influence of wave on them is ignored.

2.1 Structural model

Design parameters were detailed in a report published by the National Wind Technology Center (NWTC), proposed by Jonkman et al. [36]. Besides, the main parameters of NREL 5 MW OWT are shown in **Table 1**.

The wall thickness of the tower decreases linearly with height; the outer diameter of the tower base is 6 m and the wall thickness is 27 mm; the diameter of the tower top connecting the engine room connecting flange is 3.87 m and the wall thickness is 19 mm. A kind of A709 circular section high-strength steel with a density of

Parameters	Numerical value
Rated power/MW	5
Tower height/m	87.6
Diameter of impeller/m	126
Rated wind speed/m·s^{-1}	11.4
Tower quality/kg	347, 460
Cabin quality/kg	240, 000
Wheel and blade quality/kg	111, 000
1st natural frequency fore-aft of the tower/Hz	0.32
1st natural frequency lateral of the tower/Hz	0.31
2nd natural frequency fore-aft of the tower/Hz	2.90
2nd natural frequency lateral of the tower/Hz	2.93

Table 1.
Main parameters of wind turbine.

7850 kg/m^3, Young's modulus of 210 GPa, Poisson's ratio of 0.3 and material yield limit of 380 MPa is chosen as the material of tower. So as to take into account the influence of structural paint, flanges, bolts and welding point masses, the calculated density has been modified to 8500 kg/m^3 [37]. The structure model and parameters are shown in **Figure 1**.

Although more accurate structural dynamics results can be obtained by using FEM, the FEM method requires high computational resources, so almost all studies ignore or simplify the wind wheel and engine room as the bulk mass. Using a similar simplification, Lavassas et al. [38] studied the structural characteristics of 1.0 MW wind turbine tower under earthquake action, in which the seismic load referred to the relevant Greek seismic codes and guidelines. Hacıefendioğlu [39] used a similar simplified model to study the seismic dynamic characteristics of a 3.0 MW offshore wind turbine under shutdown condition, considering SSI effect. Ma et al. [20] calculated the structural response and load distribution characteristics of the concrete tower of 5.0 MW wind turbine under six different earthquake actions by ABAQUS using a simplified model without considering wind turbines at all.

2.2 SSI modeling

In this paper, the current Wenkel SSI model of the American Petroleum Institute (API) is adopted to consider the nonlinearity of the lateral and longitudinal soil stiffness [40]. Meanwhile, the SSI model is represented by a nonlinear spring, as shown in **Figure 2**. The calculated monopile and jacket *p-y* curves are illustrated in **Figure 3**.

The burial depth of two structures is 25 m and the burial part is divided into 5 sections, starting from 1.25 m below the seabed, with an interval of 5 m at each calculation step. The nonlinear stiffness of the soil K_n is based on piecewise calculation, the outside diameter of the jacket pile and the monopile is 2 m and 6 m respectively.

The ultimate bearing capacity of sand varies in depths. The ultimate bearing capacity of shallow and deep sand below the seabed is denoted as Eq. (1) and Eq. (2).

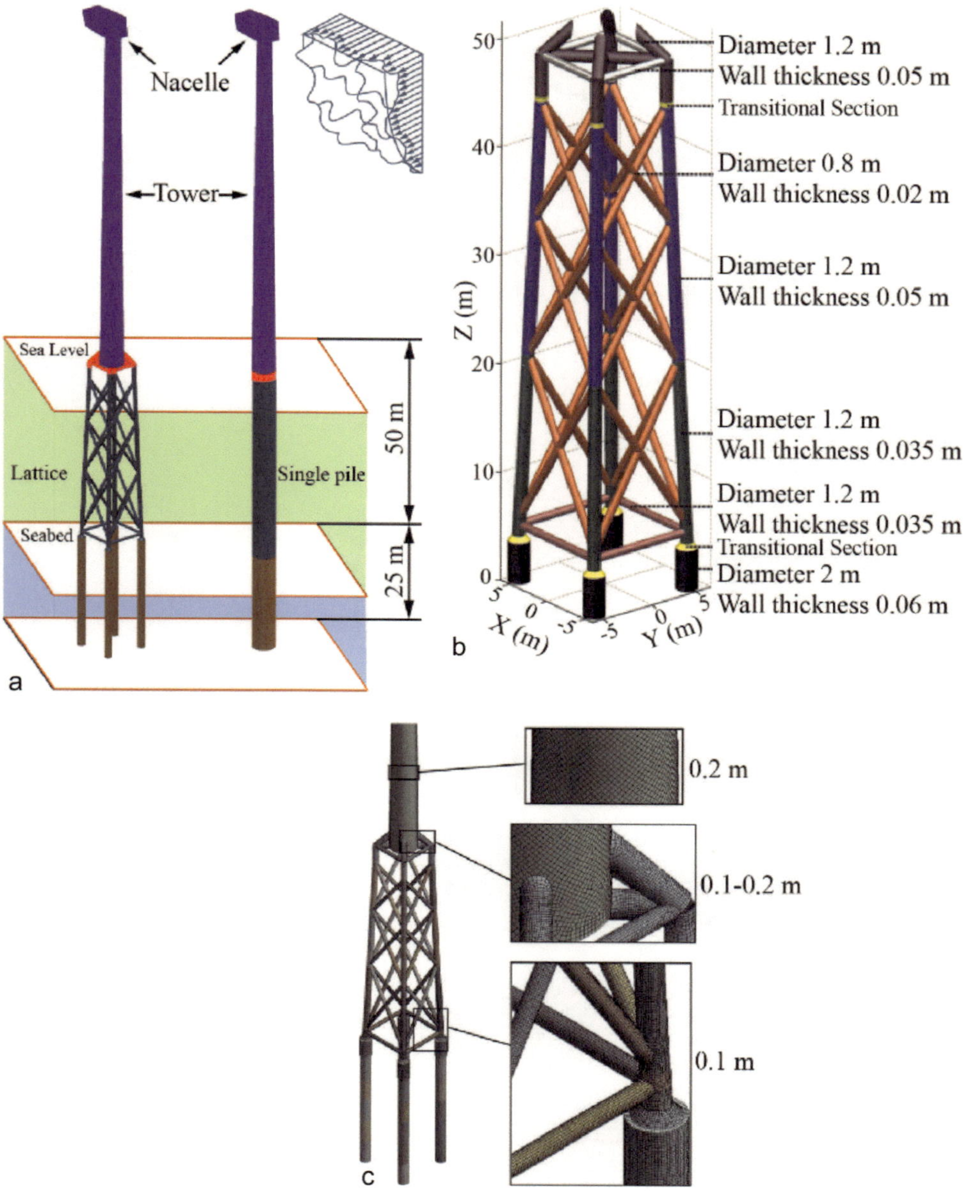

Figure 1.
Structure model and parameters. (a) Wind turbine support structure model diagram, (b) jacket model parameters and (c) jacket finite element model and grid.

$$p_s = (C_1H + C_2D)\gamma H \tag{1}$$

$$p_d = C_3D\gamma H \tag{2}$$

where p_s and p_d are the ultimate bearing capacity of shallow and deep sand respectively; γ is the effective gravity of sand; H is the depth of sand below the seabed; Coefficient C_1, C_2 and C_3 is determined by the internal friction angle φ, displayed in **Table 2**; D is the outside diameter of the pile foundation.

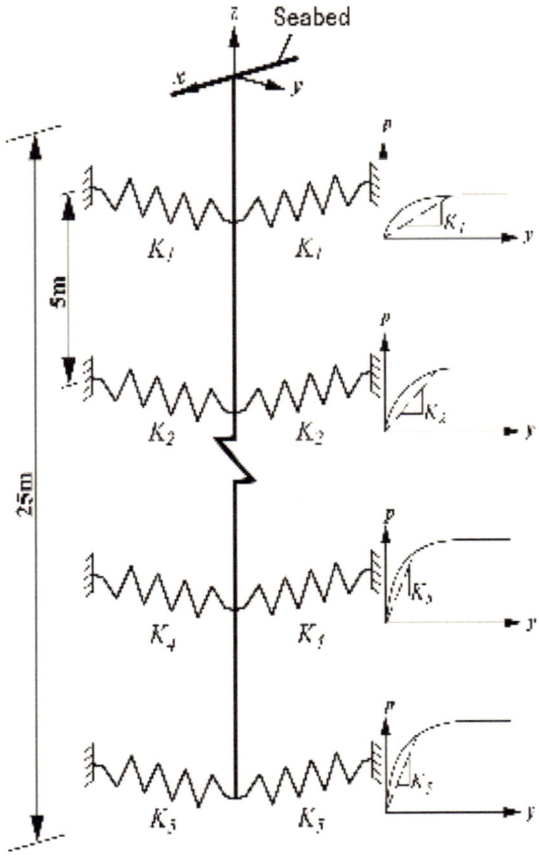

Figure 2.
Schematic diagram of soil structure interaction.

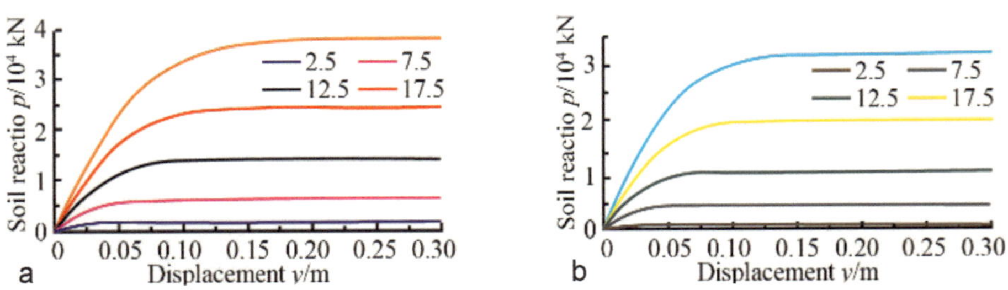

Figure 3.
p-y *curve. (a) Single pile structure and (b) jacket structure.*

When the pile foundation is laterally displaced, the sand reacts to the pile foundation due to the deformation of the pile. The nonlinear relation between the force and the lateral displacement of the sand is denoted as follows:

$$\begin{cases} p = Ap_{\mathrm{s}} \tanh\left[\dfrac{kH}{Ap_{\mathrm{s}}} y\right] \\ p = Ap_{\mathrm{d}} \tanh\left[\dfrac{kH}{Ap_{\mathrm{d}}} y\right] \end{cases} \tag{3}$$

Soil parameters/unit	Numerical value
Internal friction angle $\varphi/°$	36
C_1 coefficient/−	3.2
C_2 coefficient/−	3.6
C_3 coefficient/−	60
Effective soil severity $\gamma/(kN \cdot m^{-3})$	20
Ground reaction coefficient $k/(kN \cdot m^{-3})$	24,440

Table 2.
Soil parameters.

where A is the empirical adjustment coefficient with regard to different loads; A is 0.9 with cyclic load and $A = (3.0 - 0.8H/D)$. When with static load, k is the ground reaction coefficient; p is the sand reaction force and y is the lateral displacement.

3. Loading calculation

3.1 Turbulent wind load

According to the IEC Kaimal turbulence wind spectrum model defined by the classic IEC61400–1 [41], the three component calculations of the wind ($K = u, v, w$)are determined by:

$$S(f) = \frac{4\sigma^2 L/\overline{u}_{hub}}{(1 + 6fL/\overline{u}_{hub})^{5/3}} \tag{4}$$

where f is the frequency; $\overline{u}_{hub}$ is the mean wind speed at the hub height; σ is the standard deviation of the wind speed and L is the integral scale parameter of each velocity component. Wind field calculations are arranged at 11 grid points in the y and z directions.

The turbulent component $V(t)$ is calculated by applying an Inverse Fast Fourier Transfer (IFFT) to the Kaimal turbulent spectrum described by Power Law wind profile and Logarithmic wind profile as follows:

$$\begin{cases} V(z) = V(z_{hub})(z/z_{hub})^{0.3} \\ V(y) = V(y_{hub})\left(\ln\frac{y}{z_0} - \psi\right)/\left(\ln\frac{z_{hub}}{z_0} - \psi\right) \end{cases} \tag{5}$$

where $V(z)$ and $V(y)$ represent the wind velocity in the lateral and vertical directions; $V(z_{hub})$ is the mean velocity at the hub height z_{hub}; The value of $V(z_{hub})$ is selected as 12.0 m/s equal to the rated wind speed; z_0 is land surface roughness with a value of 0.021; ψ is vertical stability of dimensionless function.

TurbSim [42] developed by NREL is adopted to simulate the full-field turbulent wind. The generated wind field is presented in **Figure 4**. The time-varying wind speed has a peak value of over 20 m/s at the hub and an average magnitude value of 11.4 m/s as expected. The variation of wind speed is irregular in time domain and non-uniform

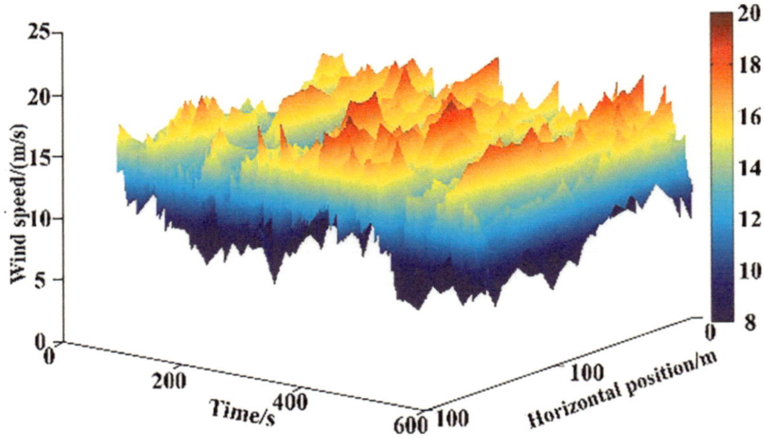

Figure 4.
Turbulent wind at hub height of the OWTs.

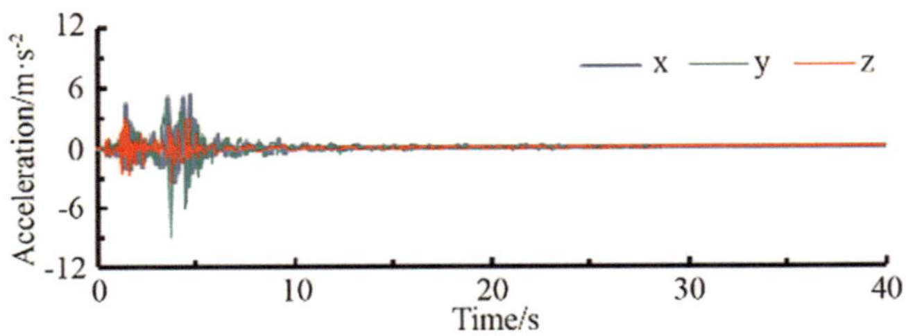

Figure 5.
Time history accelerations of ground motions.

in spatial distribution, which indicates the turbulent characteristics of the generated wind field.

3.2 Seismic load

The earthquake record was selected from the Pacific Earthquake Engineering Research Center (PEER) NGA database [43]. The earthquake event occurred in Morgan Hill, USA, with a magnitude of 6.19 and a peak acceleration of 2.44 m/s^{-2}. The time history of seismic acceleration is shown in **Figure 5**.

4. Finite element method

Transient dynamics analysis is used to analyze the dynamic response of a load structure subjected to any time variation. The Newmark time integral method is adopted to solve the whole structural system matrix for all kinds of large nonlinear elastic–plastic deformation structures [44, 45].

The basic equation of structural transient dynamics analysis is:

$$[M]\{\ddot{u}\} + [C]\{\dot{u}\} + [K]\{u\} = \{F(t)\} \tag{6}$$

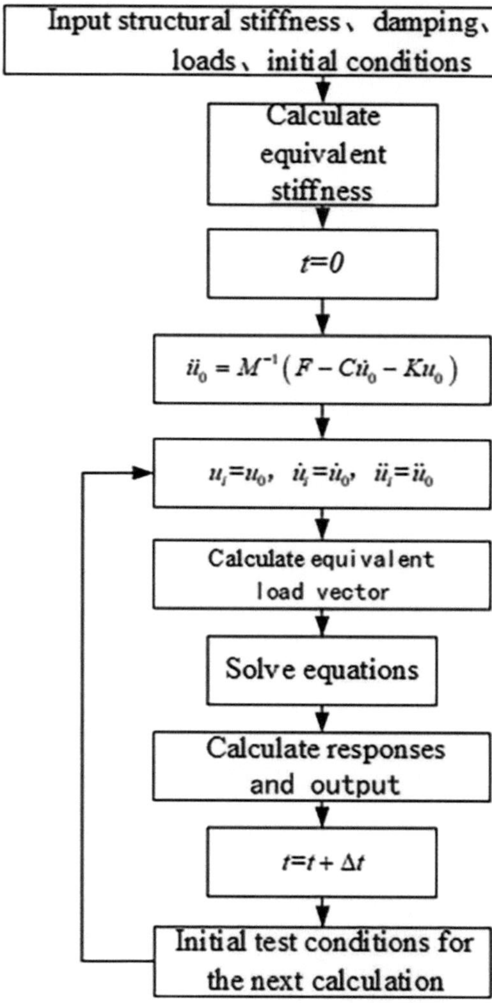

Figure 6.
Resolving flow chart of transient dynamics direct method.

where $[M]$ is mass matrix; $[C]$ is damping matrix; $[K]$ is stiffness matrix; $\{\ddot{u}\}$ is acceleration vector; $\{\dot{u}\}$ is velocity vector; $\{u\}$ is displacement vector; $\{F(t)\}$ is variable load vector.

The accuracy of the calculation result is improved by using the complete Newmark time integration method. The specific steps are as shown in **Figure 6**.

5. Reliability compliance

Like any other numerical models, finite element models should be validated before the formal application. Only in this way can we ensure the accuracy of material models, element formulations and mathematical calculations. In this study, the calculation results are compared with those of NREL wind turbine aerodyn-servodyn-hydrodyn-elasdyn open source software FAST. The dynamic response of tower-top displacement is shown in **Figure 7**.

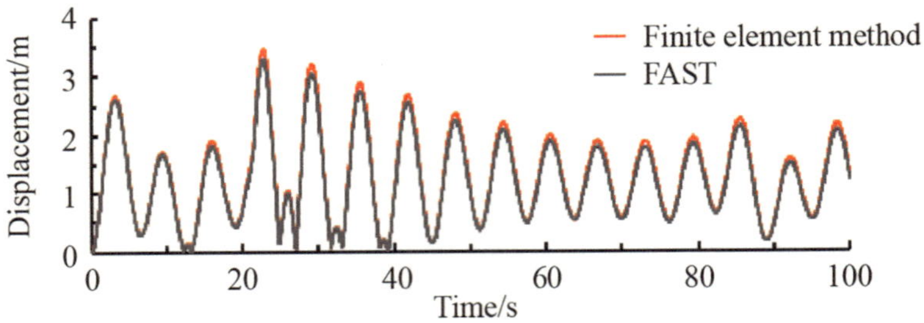

Figure 7.
Time-domain response of tower top displacement under seismic.

The result from finite element calculation is basically consistent with the result of FAST in the turbulent wind load before 20 s. Both of the curves manifest severe fluctuations, and the trend is highly consistent, with only minor differences in amplitude. The average tower-top displacement is 1.295 m via FAST, but it is 1.234 m using finite element merely with an error value of 4.7%. Therefore, the finite element model is validated.

6. Results and discussion

6.1 Modal analysis

In the modal analysis module, Block Lanczos solver performance output is used to extract mode shapes and natural frequencies. The Block Lanczos method uses the sparse matrix solver, overriding any specified solver via the Eqslv command, and can be used for large symmetric eigenvalue problems. The 1st and 2nd modal natural vibration frequencies of the two structures involved in SSI are shown in **Table 3**. The 1st and 2nd fore-aft tower vibration mode shapes are illustrated in **Figure 8**.

The vibration characteristics of the OWTs are greatly changed due to the different structures. The 1st natural frequency of the jacket structure is higher, and the 2nd natural frequency is lower, compared with those of the monopile structure. Also, the largest displacement location of the 1st mode at the nacelle is different from that of the 2nd mode.

6.2 Structural dynamic response analysis

To investigate the effect of wind-seismic interaction on the dynamic responses of the OWTs, a nonlinear time-history analysis is more appropriate compared to the

Modal	Single pile structure	Jacket structure	Amplification
Fore-aft 1st	0.157	0.180	+14.6%
Lateral 1st	0.157	0.181	+15.3%
Fore-aft 2nd	0.979	0.913	−6.7%
Lateral 2nd	0.989	0.922	−6.8%

Table 3.
Wind turbine natural frequency.

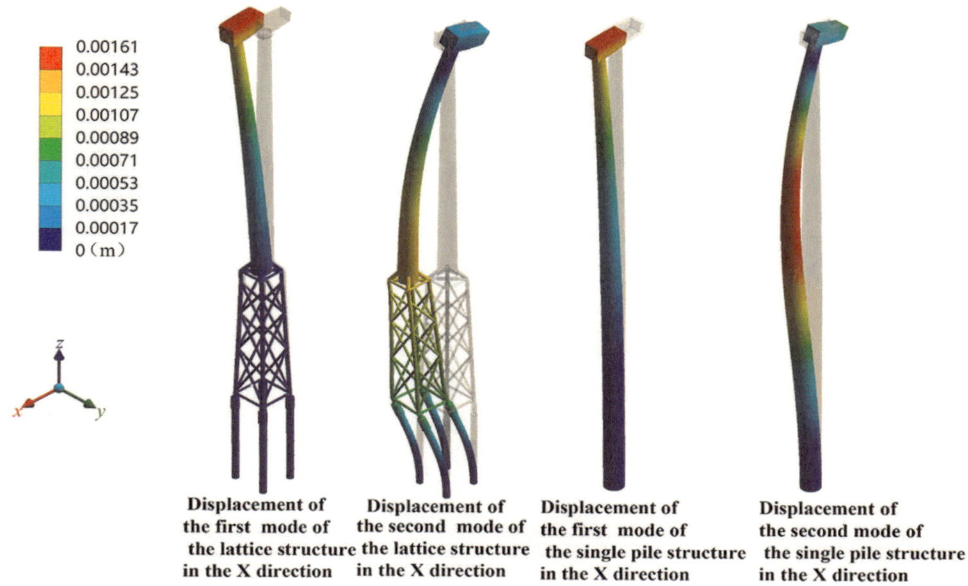

Figure 8.
Fore-aft mode shape of tower.

other analysis schemes. In this study, the dynamic responses of two structures are analyzed under two conditions, with and without seismic load.

Numerical simulations in four diverse cases are carried out:

1. Jacket OWTs under wind and seismic load;

2. Monopile OWTs under wind and seismic load;

3. Jacket OWTs under wind load merely;

4. Monopile OWTs under wind load merely.

Under such circumstance, the simulation duration is 100 s and the time step is 0.002 s. The turbulent wind speed is 11.4 m/s and the earthquake is assumed to occur at 20th s with a magnitude of 6.19. Tower-top displacements in x directions (fore-aft) and y (side-to-side) are presented in **Figure 9**.

When comparing the structure behaviors in different directions, the tower-top displacement in x direction is obviously larger than that in y direction, this is because the x direction suffers from wind load. The displacement curve varies significantly in earthquake, and it is obvious larger in y direction than that in x direction, due to the damping of wind load.

When comparing the different structure behaviors, the dynamic response of the jacket during initial period of the earthquake was smaller than that of monopole, with a 32.3% decrease in peak.

After the earthquake, energy is quickly dissipated in the turbulent wind, the tower-top displacement curve of jacket drops sharply in x direction, with a 39.2% decrease in peak compared with monopile. Most of the energy can be dissipated through the vibration of the jacket tower, which shows that the jacket can be used in aseismic design.

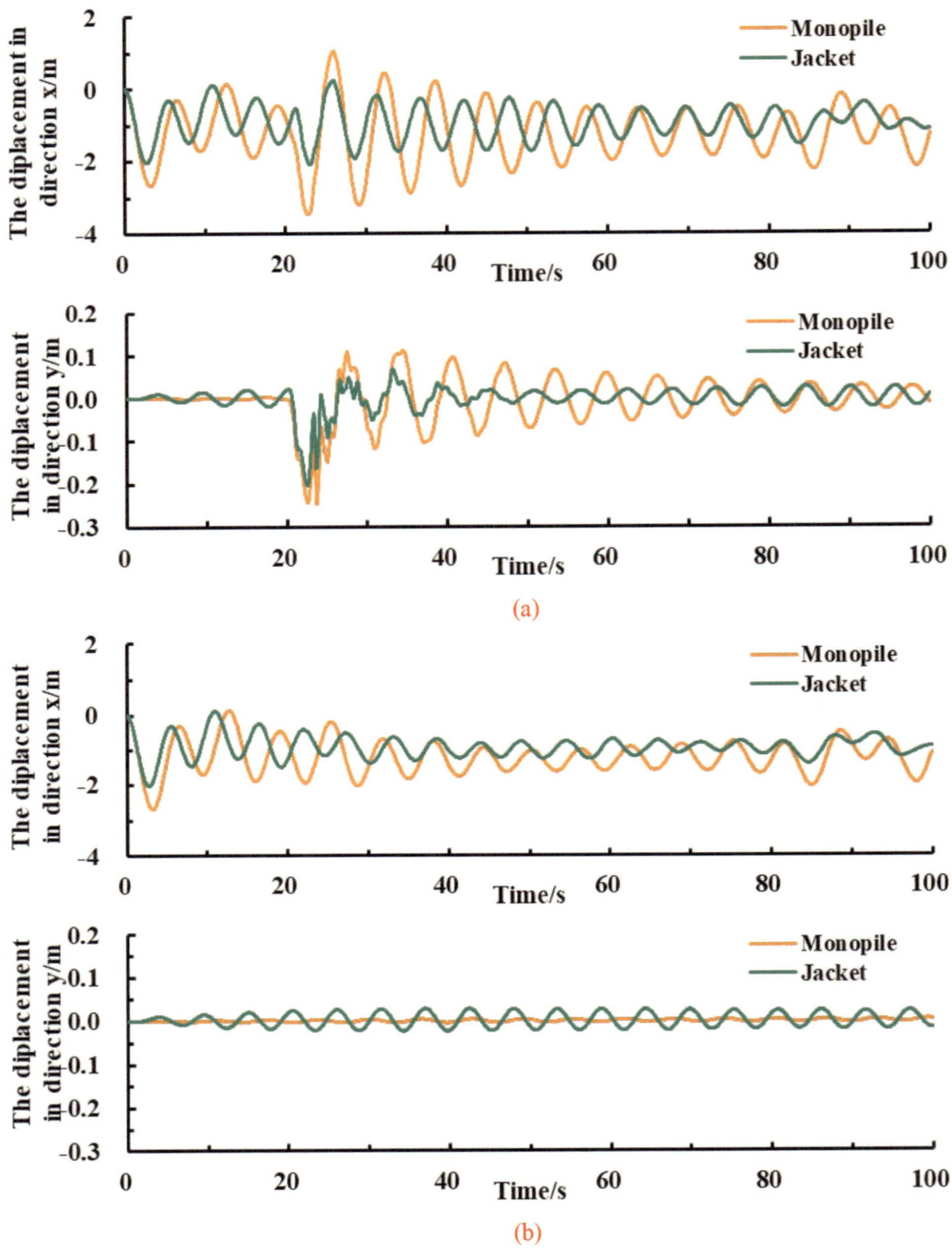

Figure 9.
Tower top displacement response in time domain. (a) Earthquake and turbulent wind combined action conditions and (b) turbulent wind action.

However, the stress of the supporting structure of OWTs will increase sharply when the earthquake occurs, and it is necessary to conduct further research on the response characteristics of shear stress. The average equivalent stress curve of the structures in time-domain is shown in **Figure 10**.

It is shown that the average equivalent stress manifests low-frequency fluctuations as wind speed vary, with a range of 0.92 ~ 64.27 MPa for monopile and 0.66 ~ 53.31 MPa for jacket. The average value of equivalent stress of jacket decreased

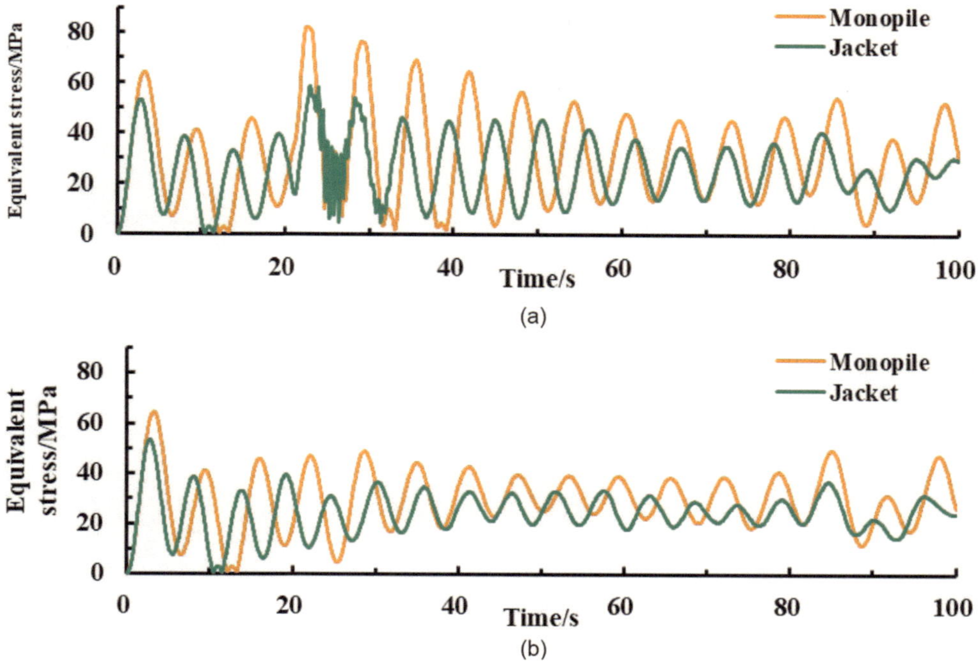

Figure 10.
Average equivalent stress curve in time domain. (a) Earthquake and turbulent wind combined action conditions and (b) turbulent wind action.

by 21.6% compared with monopile. It is indicated that the jacket structure has a higher and more reliable ultimate bearing capacity of support structures subjected to accumulated fatigue damage.

When seismic load is added, the shear curve shows high frequency fluctuations, accompanied by severe vibration of the tower. The equivalent stress amplitude varies from 1.91 to 82.05 MPa for monopile, and from 5.13 to 59.03 MPa for jacket. The average value of equivalent stress of jacket decreased by 48.7% compared with monopile. Therefore, the jacket structure can reduce the collapse possibility in that increasing rapid stress presumably exceed the tower yield limit when the earthquake hits the coast.

The principle of virtual work states that a virtual (very small) change of the internal strain energy must be offset by an identical change in external work due to the applied loads, and then the virtual strain energy value related to the nodal displacements can be got. The maximum strain energy position of the structure is shown in **Figures 11** and **12** is the cloud picture of strain energy peak moment.

The strain energy fluctuation has been proven to be associated with the change of wind speed, with an average of 228.7 J for monopile and 365.5 J for jacket. Both the two structures receive the same energy from the blades of OWT. However, due to the large displacement response of the monopile, part of the energy is dissipated. The stain energy of stable jacket is 59% higher than that of monopile.

When the earthquake occurred, the local strain energy surged in both structures. The maximum strain energy appeared at 20 s $\sim$ 30 s, the average value of jacket is 1510.8 J, which is 313% higher than that before the earthquake, and the monopile is 708.9 J, which is 94.0% higher than that before the earthquake. Compared with monopile, strain energy of the jacket is larger and more concentrated, which is prone to suffer local deformation.

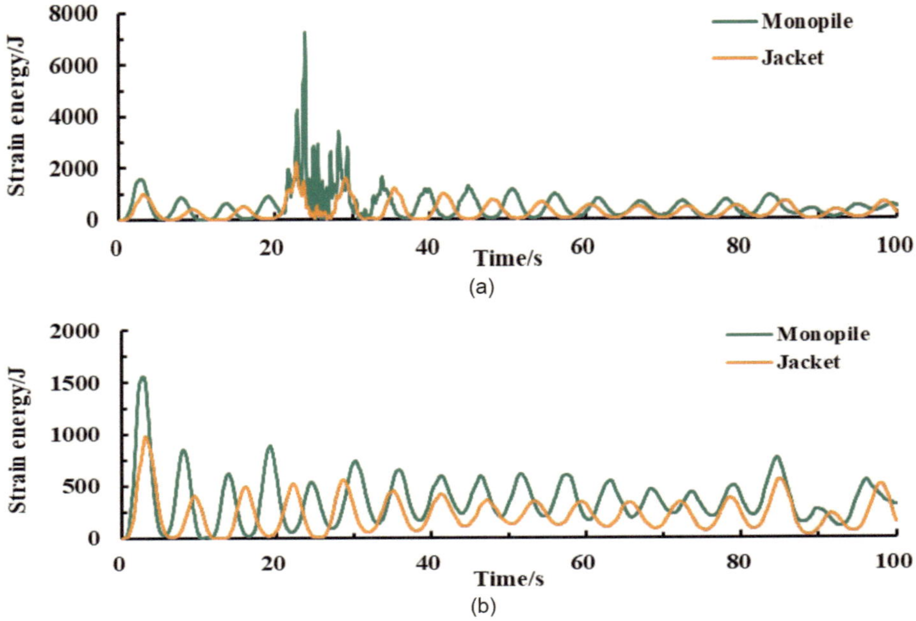

Figure 11.
Maximum strain energy position curve in time domain. (a) Earthquake and turbulent wind combined action conditions and (b) turbulent wind action.

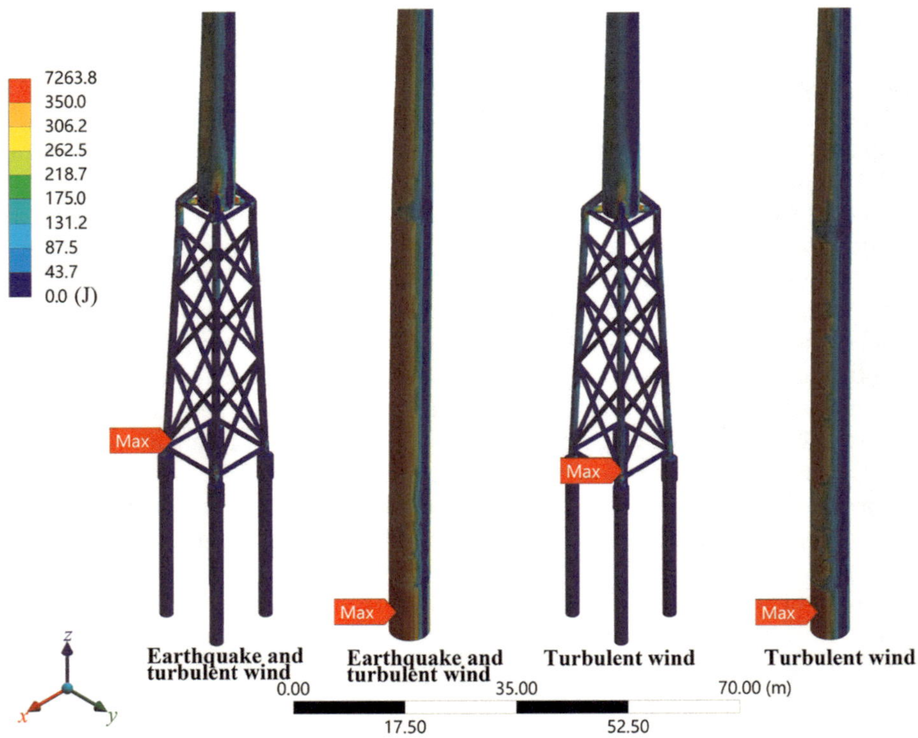

Figure 12.
Strain energy of OWT foundation structure.

7. Conclusion

In this study, the NREL 5 MW OWT was chosen as the study object and an earthquake with a magnitude of 6.19 is selected as input ground motions. Transient dynamics of large-scale OWTs under combined loads with different foundation structures is analyzed. The structural dynamic response of jacket foundation and that of monopile foundation was carefully compared. In view of the calculations and discussions described, the four key conclusions are as follows:

a. Compared with monopile-type OWT, jacket-type OWT has higher 1st natural frequency and lower 2nd natural frequency. The 2nd modal response of jacket-type OWT is smaller.

b. The tower-top displacement of the jacket-type OWT is smaller under the action of turbulent wind, which can guarantee the stability of a multitude of important components at the top of the tower. Compared with the monopile, the equivalent stress of jacket has a low average value and a small fluctuation range, which improves the structural safety.

c. When the earthquake occurs, the monopile response is more violent and lasts for a longer time, while the jacket response is moderate, and can be fully restored to the stationary state more quickly. Thus, the jacket performs better in anti-seismic behaviors.

d. The earthquake can cause the jacket-type OWT at the connection point between the jacket structure and the tower, while the monopile-type OWT deformation occurs in the deep buried location. Therefore, the maintenance cost of jacket is lower, which is beneficial to reduce the cost of electricity.

Acknowledgements

The authors would like to acknowledge the support of National Natural Science Foundation of China (Grand No. 51976131, 52006148 and 52106262), and this research also supported by the Shanghai "Action Plan for Scientific and Technological" (Grant No. 19060502200).

Author details

Peilin Wang[1], Minnan Yue[1*], Chun Li[1,2], Yangtian Yan[1], Kailun Niu[1] and Xinyu Pei[1]

1 School of Energy and Power Engineering, University of Shanghai for Science and Technology, Shanghai, People's Republic of China

2 Shanghai Key Laboratory of Multiphase Flow and Heat Transfer in Power Engineering, Shanghai, People's Republic of China

*Address all correspondence to: ymn@usst.edu.cn

References

[1] Li Y, Sun C, Yu J, et al. Scaling method of the rotating blade of a wind turbine for a rime ice wind tunnel test. Energies. 2019;**12**:626-627

[2] GWEC. Global Wind Report 2021. Brussels, Belgium: GWEC; 2021

[3] Yang Y, Ye K, Li C, et al. Dynamic behavior of wind turbines influenced by aerodynamic damping and earthquake intensity. Wind Energy. 2018;**21**(5): 303-319

[4] Ding Q, Li C, Cheng S, et al. Study on TMD control on stability improvement of barge-supported floating offshore wind turbine based on the multi-island genetic algorithm. China Ocean Engineering. 2019;**33**(3):1-13

[5] Bazeos N, Hatzigeorgiou GD, Hondros ID, Karamaneas H, Karabalis DL, Beskos DE. Static, seismic and stability analyses of a prototype wind turbine steel tower. Engineering Structures. 2002;**24**(8): 1015-1025

[6] Hacıefendioğlu K. Stochastic seismic response analysis of offshore wind turbine including fluid-structure-soil interaction. The Structural Design of Tall and Special Buildings. 2012;**21**(12): 867-878

[7] Zhao X, Maisser P. Seismic response analysis of wind turbine towers including soil-structure interaction. Proceedings of the Institution of Mechanical Engineers, Part K: Journal of Multi-body Dynamics. 2006;**220**(1): 53-61

[8] Sapountzakis EJ, Dikaros IC, Kampitsis AE, Koroneou AD. Nonlinear response of wind turbines under wind and seismic excitations with soil–structure interaction. Journal of Computational and Nonlinear Dynamics. 2015;**10**(4):041007

[9] Kaynia AM. Seismic considerations in design of offshore wind turbines. Soil Dynamics and Earthquake Engineering. 2019;**124**:399-407

[10] Asareh MA, Prowell I, Volz J, Schonberg W. A computational platform for considering the effects of aerodynamic and seismic load combination for utility scale horizontal axis wind turbines. Earthquake Engineering and Engineering Vibration. 2016;**15**(1):91-102

[11] Asareh MA, Prowell I. Seismic Loading for FAST. (No. NREL/SR-5000-53872). Golden, CO, USA: National Renewable Energy Laboratory; 2011

[12] Yang Y, Bashir M, Li C, et al. Analysis of seismic behaviour of an offshore wind turbine with a flexible foundation. Ocean Engineering. 2019;**178**:215-228

[13] Yang Y, Li C, Bashir M, et al. Investigation on the sensitivity of flexible foundation models of an offshore wind turbine under earthquake loadings. Engineering Structures. 2019; **183**:756-769

[14] Ma H, Zhang D. Seismic response of a prestressed concrete wind turbine tower. International Journal of Civil Engineering. 2016;**14**(8):561-571

[15] Patil A, Jung S, Kwon OS. Structural performance of a parked wind turbine tower subjected to strong ground motions. Engineering Structures. 2016; **120**:92-102

[16] Smith V, Mahmoud H. Multihazard assessment of wind turbine towers under simultaneous application of wind,

operation, and seismic loads. Journal of Performance of Constructed Facilities. 2016;**30**(6):04016043

[17] Tesser RK, Pilla LL, Dupros F, et al. Improving the Performance of Seismic Wave Simulations With Dynamic Load Balancing. Proceedings of the 2014 22nd Euromicro International Conference on Parallel, Distributed, and Network-Based Processing. Los Alamitos, CA: IEEE Computer Society; 2014

[18] Asareh MA, Schonberg W, Volz J. Effects of seismic and aerodynamic load interaction on structural dynamic response of multi-megawatt utility scale horizontal axis wind turbines. Renewable Energy. 2016;**86**:49-58

[19] Jin X, Liu H, Ju W. Wind turbine seismic load analysis based on numerical calculation. Strojniski Vestnik/Journal of Mechanical Engineering. 2014;**60**(10): 638-648

[20] Ma H, Dongdong Z. Seismic response of a prestressed concrete wind turbine tower. International Journal of Civil Engineering. 2016;**14**:561-571

[21] Gao Z, Saha N, Moan T, Amdahl J. Dynamic Analysis of Offshore Fixed Wind Turbines Under Wind and Wave Loads Using Alternative Computer Codes. Crete, Greece: Proceedings of the 3rd EAWE conference, TORQUE 2010: the science of making torque from wind; 2010

[22] Voormeeren SN, van der Valk PLC, Nortier BP, et al. Accurate and efficient modeling of complex offshore wind turbine support structures using augmented superelements. Wind Energy. 2014;**17**:35-54

[23] Banerjee A, Chakraborty T, Matsagar V. Stochastic dynamic analysis of an offshore wind turbine considering soil-structure interaction. Advances in Structural Engineering. 2014;**10**:673-687

[24] Zhang Y, Liao C, Chen J, et al. Numerical analysis of interaction between seabed and mono-pile subjected to dynamic wave loadings considering the pile rocking effect. Ocean Engineering. 2018;**155**:173-188

[25] Ye J, Wang G. Seismic dynamics of offshore breakwater on liquefiable seabed foundation. Soil Dynamics and Earthquake Engineering. 2015;**76**:86-99

[26] Ibsen LB, Liingaard M. Prototype bucket foundation for wind turbines—natural frequency estimation. DCE Technical report No. 9. Aalborg, Denmark: Department of Civil Engineering, Aalborg University; 2006

[27] Anastasopoulos I, Theofilou M. Hybrid foundation for offshore wind turbines: Environmental and seismic loading [J]. Soil Dynamics & Earthquake Engineering. 2016;**80**(1–2):192-209

[28] Galvín P, Romero A, Solís M, et al. Dynamic characterisation of wind turbine towers account for a monopile foundation and different soil conditions. Structure and Infrastructure Engineering. 2016;**13**(7):942-954

[29] Hacıefendioglu K. Stochastic seismic response analysis of offshore wind turbine including fluid-structure-soil interaction [J]. Structural Design of Tall and Special Buildings. 2012;**21**:867-878

[30] Kaynia AM. Seismic considerations in design of offshore wind turbines. Soil Dynamics and Earthquake Engineering. 2018;**124**:399-407

[31] Zhongsheng L, Yang Y, Chun L, et al. Analysis of dynamic seismic response characteristics in time-frequency domain of wind turbine

considering SSI [J]. Journal of Dynamic Engineering. 2018;**38**(7):587-593

[32] Yang Y, Chun L, Quanyong Y. Research on dynamic response of a 5 MW wind turbine tower under seismic conditions. Journal of Chinese Society of Power Engineering. 2017;**11**:83-89

[33] Shen-Haw J, Huang Y-C. Analyses of offshore wind turbine structures with soil-structure interaction under earthquakes. Ocean Engineering. 2019; **187**:106190

[34] Yang Y, Chun L, Weipao M, et al. Dynamic response of wind turbine structure under combined action of turbulent wind field and earthquake excitation [J]. Journal of Vibration and Shock. 2016;**34**(21):136-143

[35] Xiaoni Wu YH, Li Y, et al. Foundations of offshore wind turbines: A review. Renewable and Sustainable Energy Reviews. 2019;**104**:379-393

[36] Jonkman JM, Butterfield S, Musial W, Scott G. Definition of a 5 MW reference wind turbine for offshore system development in: Report No NREL/TP-500-38060. Golden: National Renewable Energy Laboratory; 2009

[37] Agbayani NA. A technical overview of ASCE/AWEARP 2011: Recommended practice for compliance of large land-based wind turbine support structures. Structures Congress. 2014;**2014**: 1759-1770

[38] Lavassas I, Nikolaidis G, Zervas P, et al. Analysis and design of the prototype of a steel 1-MW wind turbine tower. Engineering Structures. 2003; **25**(8):1097-1106

[39] Hacıefendioğlu K. Stochastic seismic response analysis of offshore wind turbine including fluid-structure-soil interaction. The Structural Design of Tall and Special Buildings. 2012;**21**(12): 867-878

[40] Recommended Practice for Planning, Designing and Constructing Fixed Offshore Platforms-Working Stress Design API. Washington: American Petroleum Institute; 2002

[41] IEC 61400–1. "Wind turbines-Part 1: Design requirements." 3rd ed. Geneva, Switzerland: International Electrotechnical Commission; 2005

[42] Jonkman BJ. TurbSim User's Guide: Version 1.50. (No. NREL/TP-500-46198). Golden, CO, United States: National Renewable Energy Laboratory (NREL); 2009

[43] Pacific Earthquake Engineering Research (PEER) ground motion database. Available from: https://peer.berkeley.edu/

[44] Clough RW. Thoughts about the origin of the finite element method. Computers and Structures. 2001;**79**: 2029-2030

[45] Clough RW. Early history of the finite element method from the view point of a pioneer. International Journal for Numerical Methods in Engineering. 2004;**60**:283-287

Chapter 2

Research on Operation Characteristics of Heater Directly Driven by Vertical Axis Wind Turbine

Tieliu Jiang, Shengwen Wang, Lidong Zhang and Zhongbin Zhang

Abstract

Direct heating of wind turbine is a form of heating production using wind energy. Because of its low requirements for wind quality, relatively simple device structure and high heating efficiency, the wind turbine-driven heating devices can take the place of traditional fossil energy for winter heating and achieve the purpose of reducing carbon emissions to a certain extent. Through the heater experimental platform constructed at Northeast Electric Power University, it is concluded that the liquid stirring heater can operate at 7 m/s, however; the efficiency is extremely low in the low-temperature environment. The permanent magnet eddy current heater must operate at the wind speed above 13 m/s, but it also can operate normally under the low-temperature condition. Combining the advantages and disadvantages of the two experimental devices and setting the vertical axis wind generator as the original motor, the heating efficiency of the two types of heating devices are analyzed under different working conditions, and then the adaptability of the wind turbine and the heating efficiency of the heating device are also studied.

Keywords: aerodynamic characteristics, vertical axis wind turbine, wind energy utilization coefficient, wind energy heating system, thermal production efficiency

1. Introduction

In addition to achieving the energy conversion, wind energy heating system can also get the thermal energy needed by users through the "wind energy-mechanical energy-thermal energy" route [1]. The kinetic energy of the natural wind is captured and converted to mechanical energy by the wind turbine firstly, and then the heater converts the mechanical energy to the desired thermal energy. Compared with the first type of energy conversion, the second form is called the wind energy direct heating system and it saves the power generation equipment (as shown in **Figure 1**)

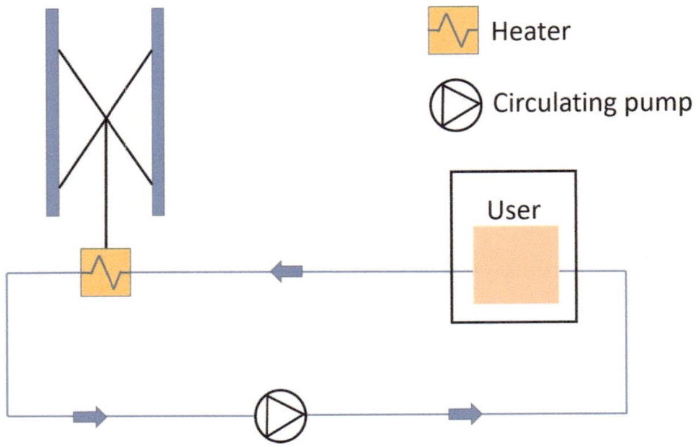

Figure 1.
Schematic diagram of the "wind energy-mechanical energy-thermal energy" conversion pathway.

and reduces the number of energy conversion time. The system will further reduce the initial investment cost and significantly improve the energy utilization coefficient.

The vertical axis wind turbine has attracted more and more attention due to its simple structure, low cost, and no yaw system required. The vertical axis wind turbine is divided into lift-type and drag-type vertical axis wind turbines, between which the lift-type one has a higher wind energy utilization coefficient under the high blade tip speed ratio, and thus the wind turbine has higher power. Lift-type vertical axis wind turbines are usually designed with two-or three-bladed wind turbines, and the three-bladed turbine has a lower shaft torque ripple and better self-starting characteristics compared to the two-bladed turbine [2].

The basic principle of the permanent magnet eddy current heater (**Figure 2**) is that when the rotor of the eddy current heater starts to rotate, the stator heating element is in the changing magnetic field. Since the stator heating element is generally a solid metal structure, many free loops can be formed inside it. Under the action of changing magnetic field, the magnetic flux of each circuit will change, which will produce an induced current. The impedance value formed in the stator heating body is small, so the circuit current will be large, so as to achieve the effect of heating.

Figure 2.
Model of the permanent magnet eddy current heater.

For the study of wind energy heating, the authors in reference [3] set up an outdoor mixing heating experimental platform powered by natural wind and calculated the heating efficiency by the recorded wind speed, rotational speed, and temperature of working fluid. The authors in reference [4] compared the heating effects of the flat blade and the cylindrical blade and found that the heating effect of the flat blade is much better. The authors in reference [5] established a mathematical model to match the torque and the power, which provides a theoretical basis for the design of the stirred wind heating device. The authors in reference [6] optimized the structure of the wind turbine by using the Fluent software according to the relevant theory of wind turbine. The authors in reference [7] used Computational Fluid Dynamics (CFD) method to analyze the thermal efficiency of wind energy heating, and verified the feasibility of using CFD to analyze the mixing heating device.

The basic principle of the liquid stirring heater (**Figure 3**) is that the wind turbine directly drives the agitator to rotate the liquid at high speed and make the liquid heat.

The authors in reference [8] studied the relationship between the torque required in the starting stage of the liquid stirring heater and the stirring impeller radius, angular acceleration. The liquid stirring heater is accompanied by a higher torque when starting.

The authors in reference [9] studied the permanent magnet eddy current heater directly driven by the vertical axis resistance differential wind turbine and analyzed the work of the permanent magnet eddy current heater under a certain wind speed. The authors in reference [10] simulated the permanent magnet eddy current heater model using the finite element method, determined the relevant geometric parameters and material properties of the model and obtained the heater power. The authors in reference [11] set up a permanent magnet eddy current heater experiment device, through which the relevant data were obtained. By using the test device, the temperature changes under different rotating speeds, working times, and different import and export water temperatures were measured, and then the corresponding conversion efficiency was obtained. The authors in reference [12] show that the increase of the thermal energy of the permanent magnet eddy current heater is roughly proportional to the square of the rotational speed increase. The authors in reference [13]

Figure 3.
Model of liquid stirring heater.

Figure 4.
Model of vertical axis wind turbine in wind tunnel test.

pointed out that for the instability and intermittent nature of wind energy, connecting the thermal energy storage device after the permanent magnet eddy current heater can ensure the stable output of thermal energy.

In this study, the operation characteristics of the heater directly driven by a vertical axis wind turbine (**Figure 4**) under different working conditions are studied. The heating efficiencies of the two types of heaters are analyzed, and the matching relationship between the wind turbine and heater is optimized (see **Figures 2–4**).

2. Physical model and numerical simulation method

2.1 Physical model of the wind turbine

The wind turbine model is a three-blade vertical axis wind turbine adopting the NACA0018 symmetrical airfoil. The string length (c) of the blade is 0.25 m, the rotating diameter (D) of the wind turbine is 0.55 m. The blade height of the vertical axis wind turbine is 1 m. The geometric model and parameters are shown in **Figure 5** and **Table 1**. The computational domain is mainly divided into two parts, the outside is the static domain and the inner is the internal rotating domain, and sliding grid calculations are applied to rotate the wind turbine in the inner rotating domain. These domains are connected by the interface connection. To eliminate the effect of the inlet blocking on the wind turbine performance and to ensure the continuity of the fluid flow during the simulation, the diameter of the rotating domain inside is set to be 1.5D (see **Figure 6**) [14].

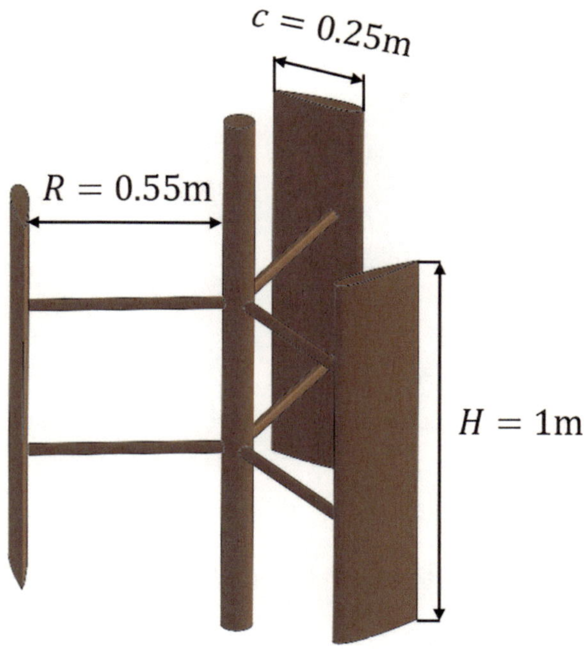

Figure 5.
Geometric model of vertical axis wind turbine.

The parameter name	Value	Unit
Blade height/H	1	m
Airfoil	NACA0018	—
Chord length/c	0.25	m
Blade number	3	—
Rotor rotating diameter/D	1.1	m

Table 1.
Dimension parameters of the vertical axis wind turbine.

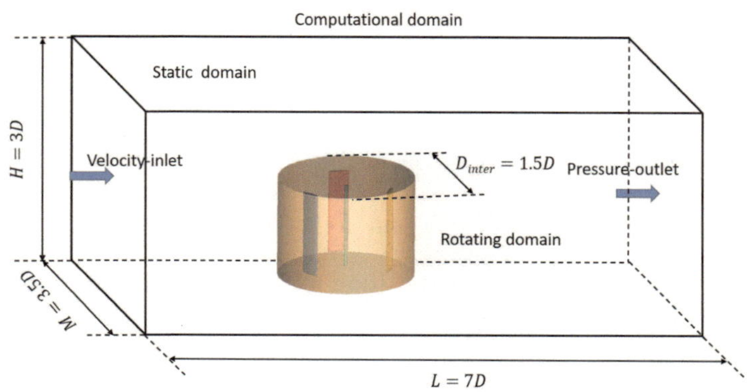

Figure 6.
Schematic diagram of the 3D Computational domain.

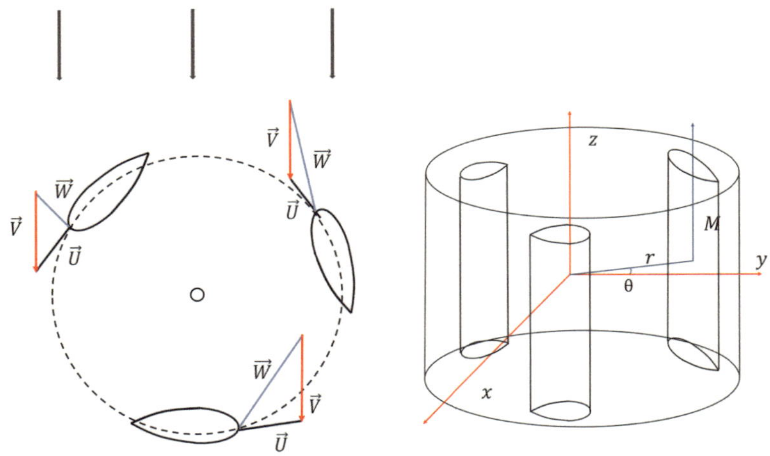

Figure 7.
Force analysis of the element on the blade.

The aerodynamic characteristics of the vertical axis wind turbine can be expressed as follows:

The following speed relation is satisfied when the wind turbine blades are in each position as Eq. (1):

$$\vec{V} = \vec{U} + \vec{W} \tag{1}$$

where, $\vec{V}$—inflow wind speed, m/s; $\vec{U}$—wind turbine blade line speed, m/s; $\vec{W}$—relative speed between airflow and blade, m/s.

The force analysis of a certain determined blade is shown in **Figure 7**. The central point of the vertical axis wind turbine is the *O* point, and the height is 2*H*. The *Oxyz* coordinate frame is defined in **Figure 7**, where *Oz* is perpendicular to the vertical rotation axis of the rotor, *Ox* is in the same direction as the wind speed passing through the rotor. And the midpoint of this element in the blade is defined as *M*, string length is *l*.

$\vec{W}$ is perpendicular to the blade direction and is expressed as Eq. (2):

$$W_n = V\sin\theta\cos\delta \tag{2}$$

The other component can be expressed as Eq. (3):

$$W_t = U + V\cos\theta = r\omega + V\cos\theta \tag{3}$$

So, the force on the blade is expressed as Eq. (4):

$$W_u^2 = W_n^2 + W_t^2 = V^2\sin^2\theta\cos^2\delta + (r\omega + V\cos\theta)^2 \tag{4}$$

Based on the above velocity decomposition relationship, the attack angle of blade is as Eq. (5):

$$\tan\alpha = \frac{V\sin\theta\cos\delta}{r\omega + V\cos\theta} \tag{5}$$

The aerodynamic pressure acting on the blade can be expressed as Eq. (6):

$$Q = \frac{1}{2}\rho W_u^2 \tag{6}$$

The Lilienthal aerodynamic coefficient at this blade element is expressed as Eq. (7):

$$\begin{cases} C_t = C_l \sin\alpha - C_d \cos\alpha \\ C_n = C_l \cos\alpha + C_d \sin\alpha \end{cases} \tag{7}$$

where C_l is the lift coefficient, and C_d is the drag coefficient.

Component forces of normal direction and wing string direction of the blade are given as Eq. (8):

$$\begin{cases} dN = C_n Qlds \\ dT = C_t Qlds \end{cases} \tag{8}$$

Decompose the above component forces to the flow wind speed direction, the resultant force received by the rotor in that direction is as Eq. (9):

$$F = \frac{bL}{2\pi}\int_{-H}^{+H}\int_{0}^{2\pi} Q(C_n \sin\theta\cos\delta - C_t \cos\theta)d\theta ds \tag{9}$$

where b is the number of blades.

The torque formula provided by the force acting on the blade for the rotor rotation axis is expressed as Eq. (10):

$$dM = dTr = C_l QLrds \tag{10}$$

Integrating the above formulas, the torque of the whole rotor is expressed as Eq. (11):

$$M = \frac{bl}{2\pi}\int_{-H}^{+H}\int_{0}^{2\pi} C_l Qrdsd\theta \tag{11}$$

Therefore, the power is expressed as Eq. (12):

$$P = M\omega = \frac{bl}{2\pi}\int_{-H}^{+H}\int_{0}^{2\pi} C_l Qr\omega dsd\theta \tag{12}$$

The wind energy utilization coefficient C_P is expressed as Eq. (13):

$$C_P = \frac{P}{\frac{1}{2}\rho V^3 A} \tag{13}$$

2.2 Grid division and verification of numerical simulation results

Grid division is a very important part in numerical simulation. Good grid division can improve the accuracy of the wind turbine performance prediction. In order to ensure the accuracy of the simulation, this study adopts the 3D structured grid for the wind turbine as shown in **Figure 8**.

For the three-blade wind turbine model studied in this paper, the pressure-velocity coupling method and SIMPLE algorithm are used to solve the transient URANS Equation, and the pressure order, momentum term, and turbulence dissipation term

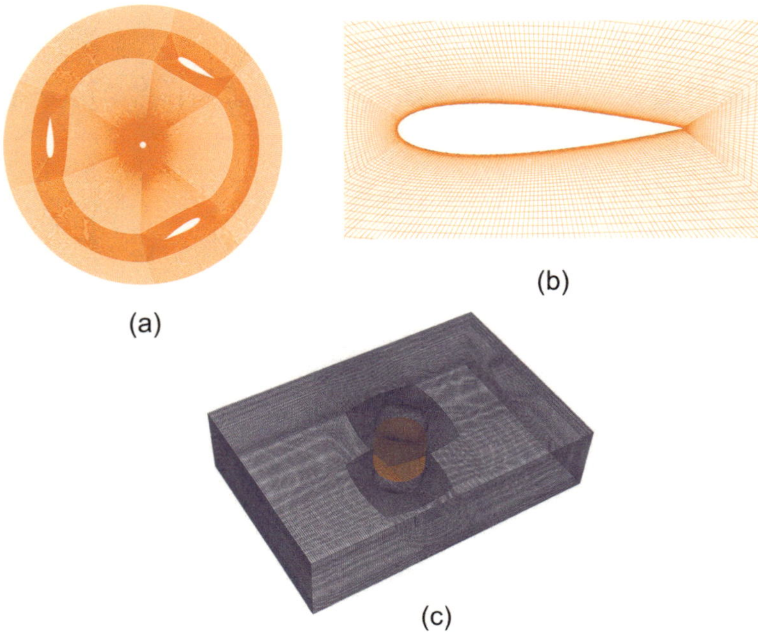

Figure 8.
Mesh of vertical axis wind turbine: (a) top view of mesh in rotating domain; (b) mesh around blade; (c) the entire mesh in the computational domain.

are all solved by the second-order windward space dissipation algorithm, and the judgment criterion of convergence is set to 10^{-5}. The turbulence model used in the simulation is the transition SST turbulence model, which is the original *k*-*ω*, SST Equations plus two empirical formulas. It is more accurate in the turbulence rotation compared to other simulated wind turbine models. In order to avoid the influences of the grid number, the y^+ value, the turbulence model, the algorithm, the time step, and other factors on the simulation results, it needs to verify the simulation results, as shown in **Figure 9**. Compared with the 2D numerical simulation results, it can be

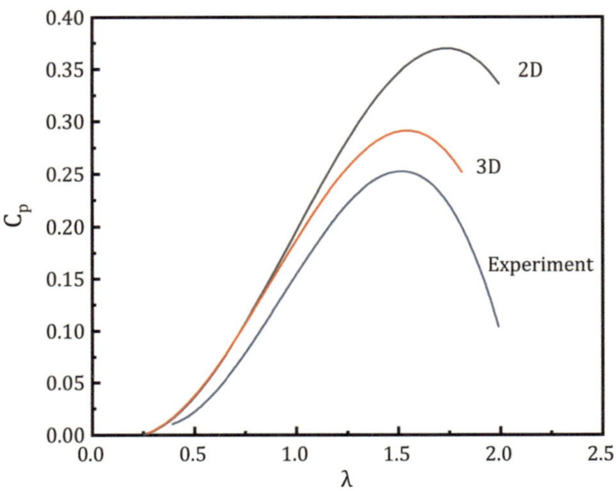

Figure 9.
Comparison of 3D numerical simulation results to experimental data and 2D numerical simulation results.

found that the wind energy utilization coefficient using the 3D numerical calculation results is close to the change trend of the wind tunnel experiment, and the accuracy of the wind energy utilization coefficient prediction is obviously better compared with the 2D numerical simulation. When the wind turbine runs at the optimal tip speed ratio (λ = 1.48), the wind energy utilization coefficient obtained from the 3D simulation is 0.297, which is 15% higher than that in the wind tunnel experiment and 19% lower than that in the 2D simulation.

3. Results and discussion

3.1 Operation characteristics and efficiency of the permanent magnet eddy current heater

It is known from experiments that when the working fluid temperature of the bulk heat exchange surface increases by 60°C under different speed conditions, the faster the speed of the permanent magnet eddy current heater, the faster the working fluid temperature of the heat exchanger surface increases. However, if the speed is too fast, it will lead to an increase in heat production absorbed by the bulk metal during the heater operation, the uniform heating of circulating working fluid, and a high energy loss. Overall, the running speed of the heater is relatively suitable at 20 rad/s. The starting torque of the permanent magnet eddy current heater is about 6.89N·m, while after the magnetic eddy current heater is used as the wind turbine load, the wind turbine tip speed is relatively low, which will deviate from the optimal tip speed ratio interval, and thus leads to a low wind energy utilization coefficient during operation.

For the permanent magnet eddy current heater, when the wind speed varies from 13 m/s to 17 m/s, the output parameters of the wind turbine are shown in **Table 2**.

The rotating speed and torque curve of wind turbine under different wind speed conditions are shown in **Figures 10** and **11**. Clearly, the rotating speed and torque of the wind turbine both increase linearly with the wind speed increased.

Wind speed/ (m/s)	Rotating speed/ (rad/s)	Tip speed ratio	C_P	Torque/ (N m)	Power of wind turbine/(W)
13.0	5.76	0.24	0.027	7.01	40.39
13.5	6.80	0.27	0.033	8.09	55.09
14.0	8.37	0.33	0.044	9.61	80.54
14.5	9.94	0.38	0.045	11.33	112.78
15.0	11.51	0.42	0.067	13.23	152.45
15.5	13.08	0.46	0.080	15.24	199.60
16.0	15.70	0.59	0.127	16.70	350.34
16.5	17.79	0.60	0.128	21.54	383.52
17.0	19.89	0.64	0.150	24.89	495.21

Table 2.
Output parameters of wind turbine at 13–17 m/s wind speeds.

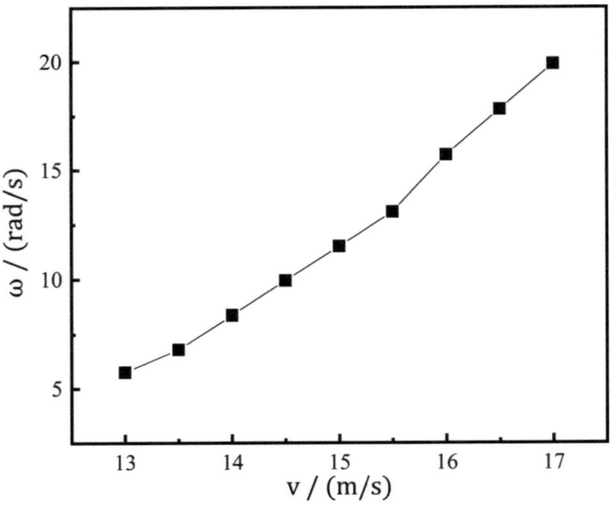

Figure 10.
Curve of the rotating speed change at 13–17 m/s wind speeds.

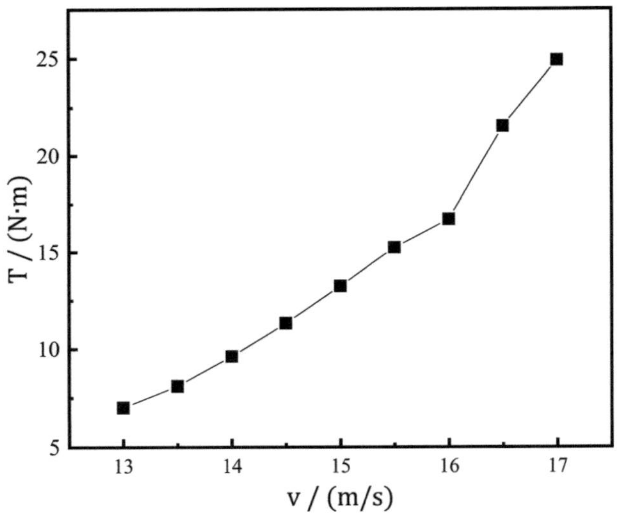

Figure 11.
Curve of the torque change at 13–17 m/s wind speeds.

According to the experimental data, the relationship between the heat absorption power P_{a1} of the permanent magnet eddy current heater during normal operation and the mechanical power P_w associated with the input heater is obtained as Eq. (14):

$$P_{a1} = 0.0008P_w^2 + 0.3201P_w + 2.875 \tag{14}$$

The relationship between the power of heat absorption P_{a2} of liquid stirring heater and the mechanical power P_w associated with the input heater is obtained as Eq. (15):

$$P_{a2} = 0.0003P_w^2 + 0.366P_w - 10.061 \tag{15}$$

Wind speed/ (m/s)	Power of wind turbine/(W)	Thermal energy exchange power/(W)	Heating efficiency/(%)	System efficiency/(%)
13.0	40.39	17.11	42.36	1.14
13.5	55.09	22.94	41.64	1.37
14.0	80.54	33.85	42.02	1.85
14.5	112.78	49.15	43.58	1.96
15.0	152.45	70.27	46.09	3.09
15.5	199.60	98.64	49.42	3.95
16.0	350.34	213.21	60.68	7.73
16.5	383.52	243.31	63.44	8.12
17.0	495.21	357.58	72.21	10.83

Table 3.
The heating efficiency and system efficiency of the permanent magnet eddy current heater.

Therefore, the efficiency of the heater can be expressed as Eq. (16):

$$\eta_{pth} = \frac{P_a}{P_m} \tag{16}$$

where P_a—heat absorption power of working fluid, J; P_m—input mechanical power of heater, J; η_{pth}—heater efficiency.

The system efficiency is the ratio of the heat obtained by the circulating working fluid to the wind energy swept by the wind turbine. It can directly reflect how much energy the heating system captures from the natural wind. The expression is Eq. (17):

$$\eta_{sys} = C_P \eta_{pth} \tag{17}$$

According to the above formulas, the heating efficiency and system efficiency of the permanent magnet eddy current heater can be obtained (**Table 3**) by using the output power of the wind turbine.

As seen from **Figures 12** and **13**, with the increase of the test wind speed, the heating efficiency increases at the same time. The heating efficiency and system efficiency are significantly increased when the wind speed is higher than 15.5 m/s. When the wind speed is 17 m/s, the heating efficiency and system efficiency reach the maximum values of 72.21% and 10.83%, respectively. Hence, the permanent magnet eddy current heater has higher efficiency under the condition of higher wind speed and rotating speed.

3.2 Operation characteristics and efficiency of the liquid stirring heater

According to the experiment results, the temperature rise rate fluctuates in a certain range when the liquid stirring heater rotates at different speeds, but it does not attenuate with the increase of the working fluid temperature. Hence, with the rise of work fluid temperature, the increase of environmental thermal dissipation will not significantly affect the working fluid temperature rise rate. The changes in

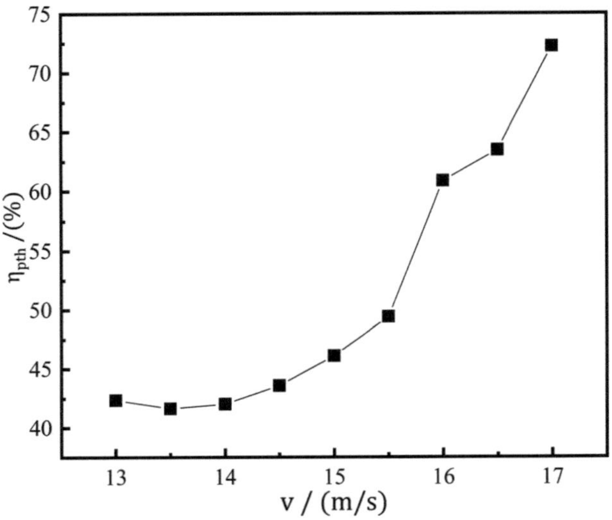

Figure 12.
Curve of the heating efficiency change at 13–17 m/s wind speeds.

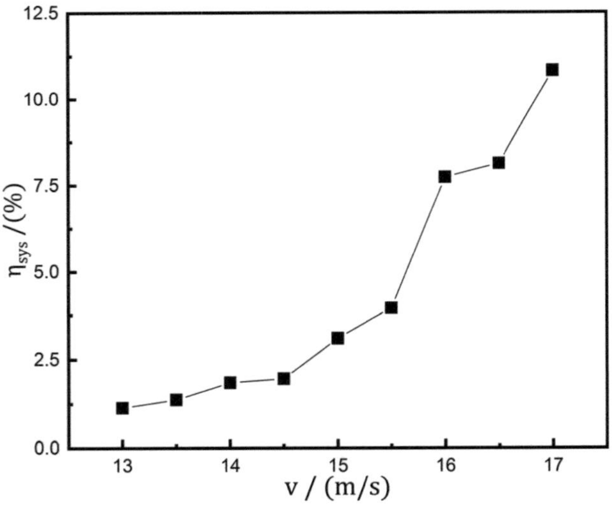

Figure 13.
Curve of the system efficiency change at 13–17 m/s wind speeds.

working fluid temperature at different heater speeds are in linear function, and the effect of heater speed on heating is significant. The strength and stiffness of mixing blade and the plate under high speed are also need to be considered.

The wind turbine can complete the start-up operation at a low wind speed, and the starting torque is much lower than the permanent magnet eddy current heater at the same power level. Therefore, compared with the permanent magnet eddy current heater, the mixing heater can use the wind energy at a lower speed and improve the wind energy utilization. Due to the small drive torque of the stirring heater, the wind turbine speed is increased, the tip speed ratio is close to the optimal tip speed ratio with a higher wind energy utilization coefficient.

Wind speed/(m/s)	Rotating speed/(rad/s)	Tip speed ratio	C_P	Torque/ (N m)	Power of wind turbine/(W)
7.0	13.61	1.07	0.202	3.62	49.30
8.0	18.84	1.29	0.264	5.10	96.17
9.0	21.98	1.34	0.281	6.63	145.75
10.0	26.17	1.43	0.292	7.94	207.76
11.0	29.30	1.46	0.295	9.53	279.37
12.0	32.44	1.49	0.297	11.26	365.15
13.0	35.59	1.51	0.295	12.96	461.13

Table 4.
Output parameters of wind turbine at 7–13 m/s wind speeds.

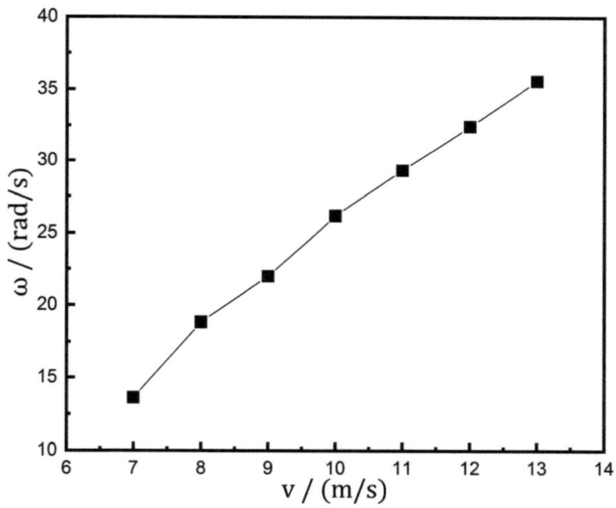

Figure 14.
Curve of the rotating speed change at 7–13 m/s wind speeds.

When the wind speed varies from 7 m/s to 13 m/s, the output parameters are given in **Table 4**.

The rotating speed and torque curve of the wind turbine in the test wind speed interval are shown in **Figures 14** and **15**. It can be seen that the rotating speed and torque both change linearly in the test wind speed range. Therefore, when the rotating speed of the heater increases, the mixing resistance is borne by the blade and the flow resistance plate will also increase simultaneously. When the heater power is necessary to be further improved, the design of the mixing blade and the damping plate structure should be optimized to increase the mixing resistance and reduce the maximum rotating speed.

Based on Eqs. (15)–(17), the heating efficiency and system efficiency (**Table 5**) of the liquid stirring heater can be obtained according to the output power of the wind turbine.

According to **Table 5**, **Figures 16** and **17**, the heating efficiency increases with the wind speed increased in the test wind speed range, and the maximum efficiency is 48% when the wind speed is 13 m/s. The corresponding wind energy utilization

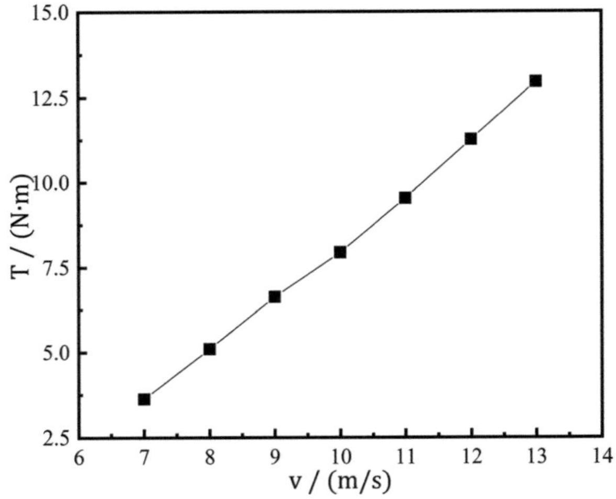

Figure 15.
Curve of the torque change at 7–13 m/s wind speeds.

Wind speed/ (m/s)	Power of wind turbine/(W)	Thermal energy exchange power/(W)	Heating efficiency/(%)	System efficiency/(%)
7.0	49.30	8.71	16.67	3.57
8.0	96.17	27.91	29.02	7.66
9.0	145.75	49.66	34.07	9.57
10.0	207.76	78.93	37.99	11.09
11.0	279.37	115.60	41.38	12.21
12.0	365.15	163.58	40.80	13.31
13.0	461.13	222.50	48.25	14.23

Table 5.
The heating efficiency and system efficiency of the liquid stirring heater.

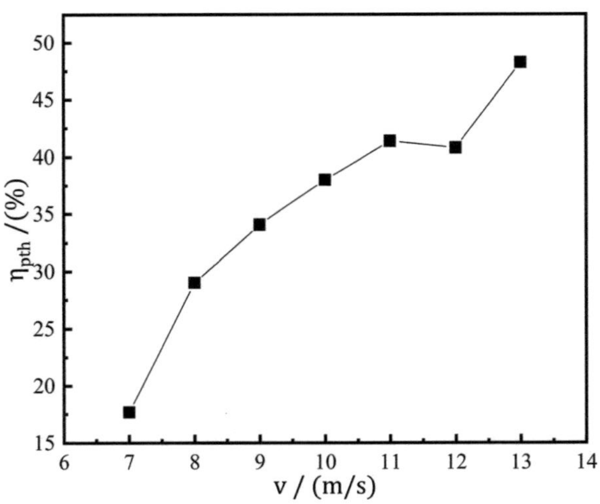

Figure 16.
Curve of the heating efficiency change at 7–13 m/s wind speeds.

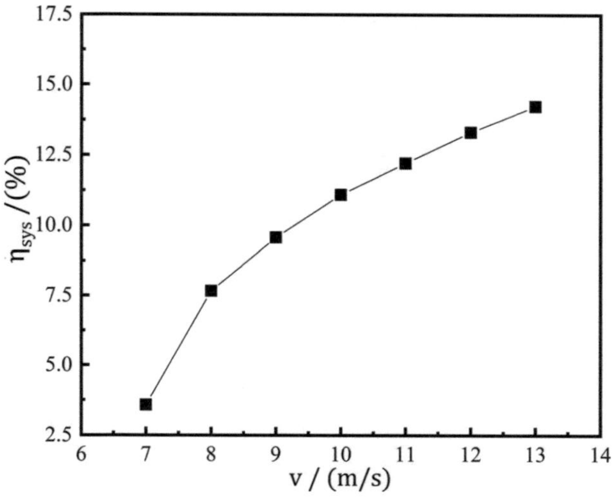

Figure 17.
Curve of the system efficiency change at 7–13 m/s wind speeds.

coefficient is 0.295 with the highest system efficiency. Though the permanent magnet eddy current heater efficiency is high, while the matching characteristics with the wind turbine are poor, leading to low system efficiency. Thus, the good matching of heater and wind turbine will effectively improve the efficiency of the wind energy heating system.

4. Conclusion

According to the existing experimental results, the numerical simulation method is applied to study the vertical axis wind turbine under different working conditions, the conclusion are as follows:

1. The input torque of the liquid stirring heater has a linear relationship with the rotating speed when its geometric structure and working fluid are determined. Therefore, replacing the working quality with high viscosity can effectively reduce the volume of the heating device. On the basis of guaranteeing the quantity of heat, optimizing the type of the stirring blade and flow resistance plate can reduce the working speed of the device, achieve a good match with the wind machine, improve the utilization coefficient of the wind turbine, and improve the heating efficiency of the system. According to the numerical simulation results, the maximum heating efficiency is up to 48.25%.

2. Compared with the liquid stirring heater in the same power level, the starting torque of the permanent magnet eddy current heater is higher, and due to the poor self-starting characteristics of the vertical axis wind turbine, the permanent magnet eddy current heater driven by the wind turbine can be put into operation at high wind speed, which cannot effectively use the wind energy at low wind speed. The wind energy utilization coefficient of the system can be improved by the cooperative operation with the liquid stirring

heater. According to the numerical simulation results, the maximum heating efficiency is up to 72.21%.

3. The heater directly driven by a vertical axis wind turbine system has a certain referred significance for other permanent magnet eddy current heater and liquid stirring heater with vertical axis wind turbine.

Author details

Tieliu Jiang*, Shengwen Wang, Lidong Zhang and Zhongbin Zhang
School of Energy and Power Engineering, Northeast Electric Power University, Jilin, China

*Address all correspondence to: jiangtieliu@163.com

References

[1] Jianfeng S. Wind energy-thermal energy direct conversion. Cereals and Oils Processing (Electronic Version). 1988;**1**:37-39. Available from: https://t.cnki.net/kcms/detail?v=8DeTHzDoIA6bkerXDuqBIgXhgCNDeatEsKvVgKyNfX40D6aYx25vSdkbssE0Z2uxg64kugubKZyrOEbWori9ubdNtSLb1kWl9Wr2t-u8wdiL0Kps8E3C3A==&uniplatform=NZKPT [Accessed: January 31, 1988]

[2] Hand B, Kelly G, Cashman A. Aerodynamic design and performance parameters of a lift-type vertical axis wind turbine: A comprehensive review. Renewable and Sustainable Energy Reviews. 2021;**139**. DOI: 10.1016/j.rser.2020.110699 [Accessed: April 1, 2021]

[3] Peng K, Yongguang L. Direct stirring heating experiment powered by the natural wind. Journal of Shanghai University of Electric Power. 2012;**28**(6):521-524. Available from: https://t.cnki.net/kcms/detail?v=8DeTHzDoIA7PwdiaE3HhKPn9DUItEtDrPRzSdSd9PCK0zLkIchJPXjEo7sPs-X4DRb-JXj2kalSK3I951-9DMbd2cpzCwTLNOPuikETvwl9CLi-UOhceRg==&uniplatform=NZKPT [Accessed: December 15, 2012]

[4] Ting G, Yongguang L, Lihua Z, et al. Experimental study on stirring thermal properties of straight and cylindrical blades. Journal of Shanghai University of Electric Power. 2015;**31**(2):156-160. Available from: https://t.cnki.net/kcms/detail?v=8DeTHzDoIA6EoFxMlJ8Y39C3aE_rTbj0dDDLXp7jjhCOIsjgG_kLvet1nQhgRXuWJnN-AmXUg7jdphvct60ciADnbHCZ6kNOh66ZiXLDMmSIkHYFCM46f9U6fIh4HGu-&uniplatform=NZKPT [Accessed: April 15, 2012]

[5] Yang L, Yihuai H. Parameter design of stirred wind heating generation device. Acta Energiae Solaris Sinica. 2014;**35**(10):1977-1980. Available from: https://kns.cnki.net/kcms/detail/detail.aspx?FileName=TYLX201410024&DbName=CJFQ2014 [Accessed: October 28, 2014]

[6] Lin M. Research on hydraulic wind energy thermal generation system [thesis]. Nanjing: Nanjing University of Science and Technology; 2015. Available from: https://kns.cnki.net/kcms/detail/detail.aspx?FileName=1016225821.nh&DbName=CMFD2017 [Accessed: December 1, 2015]

[7] Jianzhu Z, Yufeng Z, Guoye W, et al. Study on the thermal efficiency of layered liquid stirring heating system. Acta Energiae Solaris Sinica. 2014;**35**(6):1034-1039. Available from: https://kns.cnki.net/kcms/detail/detail.aspx?FileName=TYLX201406019&DbName=CJFQ2014 [Accessed: June 28, 2014]

[8] Yongguang L, Yunling M. Study on start-up performance of stirring wind thermal device. Acta Energiae Solaris Sinica. 2020;**41**(3):29-33. Available from: https://kns.cnki.net/kcms/detail/detail.aspx?FileName=TYLX202003005&DbName=DKFX2020 [Accessed: March 28, 2020]

[9] Fubao W, Maosheng Z, Haipeng T. Analysis of heating generation of vertical axial resistance. Journal of Northwest University (Natural Science Edition). 2013;**43**(4):545-548. Available from: https://kns.cnki.net/kcms/detail/detail.aspx?FileName=XBDZ201304012&DbName=CJFQ2013 [Accessed: August 25, 2013]

[10] Nebi O, Fireţeanu V. Finite element electromagnetic 2D model of an eddy current heater with rotating permanent magnets. Annals of the University of

Craiova Romania Series: Electrical Engineering. 2008;**32**(2):62-67. Available from: https://elth.ucv.ro/fisiere/anale/2008/en_2008_2.htm [Accessed: October 9, 2008]

[11] Sobor I, Rachier V, Chiciuc A, et al. Small wind energy system with permanent magnet eddy current heater. Telecommunications Energy Conferences Mart Power and Efficiency. 2013;**6**(4):143-150. Available from: http://repository.utm.md/handle/5014/14856 [Accessed: November 28, 2013]

[12] Tudorache T, Melcescu L. Outer rotor eddy current heater for wind turbines. Renewable Energy and Environmental Sustainability. 2016;**1**:25-27. DOI: 10.1051/rees/2016026 [Accessed: June 17, 2016]

[13] Zhaozheng W, Yujun G, Xiaohui Z, et al. Research and development of permanent magnet eddy current heater driven by wind energy. Energy Conservation. 2021;**40**(04):25-28. Available from: https://kns.cnki.net/kcms/detail/detail.aspx?FileName=JNJN202104008&DbName=DKFX2021 [Accessed: April 25, 2021]

[14] Yan L, Tingting Z, Kotaro T, et al. Numerical simulation on torque characteristics of vertical axis wind turbine with curved plate adding at blade trailing edge. Journal of Northeast Agricultural University. 2017;**48**(3):72-79. Available from: https://kns.cnki.net/kcms/detail/detail.aspx?FileName=DBDN201703010&DbName=CJFQ2017 [Accessed: March 21, 2017]

Chapter 3

Wind Tunnel Test of Icing Distribution on the Leading Edge of a 2D Blade Airfoil for Wind Turbines

Yan Li, Zhongqiu Mu, Zhiyuan Liu, Wenfeng Guo, Fang Feng and Kotaro Tagawa

Abstract

As clean and renewable energy, wind energy has been widely used in the world. The wind turbine is a kind of rotating machinery, which can convert wind energy into mechanical energy and electrical energy. Wind turbines sometimes face a variety of extreme weather conditions, such as icing, heavy snow, lightning, sand storm, and so on, which affect the safety operation of wind turbines. In the present study, icing events on the blade surfaces of wind turbines are focused, and the wind tunnel test of icing was carried out on a 2D blade with NACA0018 airfoil used in wind turbines. In the icing tests, three kinds of ambient temperatures were selected, including −5 °C, −10 °C, and −15 °C, and two kinds of wind speeds were decided, including 5m/s and 10m/s. The icing distributions on the blade surface at the different attack angles were tested and recorded by a high-speed camera for several minutes. The ice accretion and distribution characteristics at the leading edge of the 2D blade airfoil were acquired and analyzed. The findings can provide a reference for the research on the icing mechanism and the de- and anti-icing of wind turbines.

Keywords: wind turbine, rotating machinery, blade, icing, wind tunnel test

1. Introduction

As clean and renewable energy, wind energy has been widely used in the world and will be paid more and more attention [1]. A wind turbine is the most important equipment for the utilization of wind energy, and it is also a kind of rotating machinery, which can convert wind energy into mechanical energy, heat energy, and electrical energy [2]. According to the rotation mode of a wind turbine shaft, there are two key kinds of wind turbines, which are the horizontal axis wind turbine (HAWT) and the vertical axis wind turbine (VAWT). Currently, the HAWTs are used for the large-scale wind turbines mostly. In contrast, the VAWTs are often used for the small-scale wind energy utilization [3]. Thanks to years of researching, the performance of

Figure 1.
Icing on the blades of HAWT.

wind turbines has been greatly improved. The research focus also extends from the improvement of efficiency to how to ensure the safe, stable, and efficient operation of wind turbines. Meanwhile, global climate change intensifies and extreme weather occurs more frequently. Wind turbines are facing more severe weather conditions, such as typhoon, earthquake, lightning, blizzard, icing, etc. [4]. In the cold and moist regions, wind turbines will face the icing problem. When the environment meets a certain condition, supercooled water droplets in the air will accrete on the wind turbine surface and become icing [5]. An icing photo of a wind turbine taken by us in Northeast China in December 2019 is shown in **Figure 1**.

Icing event occurring on the wind turbine blade surface will greatly degrade the efficiency and safety of operation. The research contents of icing mainly include distribution and characteristics of icing on a blade [6–8], aerodynamic performance effect of icing on blade and rotor [9–11], anti-icing and de-icing methods and technologies [12–15], icing detection technologies [16], etc. There are two main icing research methods, numerical simulation [17, 18] and icing wind tunnel test [19, 20]. In the present study, a small-scale icing wind tunnel was used to obtain the distribution characteristics of icing on a model of 2D blade with NACA0018 airfoil. Under different ambient temperatures and wind speeds, experiments on ice accretion on the 2D blade at two kinds of angel of attacks for 4 minutes were carried out. The icing distributions were recorded and analyzed. This research can provide a reference for researching icing mechanism and de- and anti-icing technologies of wind turbines.

2. Methods

2.1 Experimental system

Figure 2 shows an icing wind tunnel experimental system in Northeast Agricultural University used in this study. It's a low-speed and return-flow type wind tunnel. The refrigeration and spray devices were installed in a normal wind tunnel to provide low temperature and icing environment. The detailed information about this system can be found in the reference [21]. The test section is 250 mm × 250 mm. The range of wind speed is from 1 m/s to 20 m/s. The ambient temperature can be controlled lower to −20°C. By using different sprayers, the icing wind tunnel can supply different kinds of liquid water content (LWC) and medium volume diameter (MVD) of the supercooled water droplet. The LWC is from 0.1 g/m^3 ~ 5 g/m^3, and the MVD is from 20 μm ~ 100 μm.

2.2 Test model

Figure 3 shows the test model of blade segment with airfoil profile used in wind turbine. To obtain the basic and normal icing characteristics of wind turbine blade, the symmetrical airfoil of NACA0018 was selected because it is often used for wind turbines and basic researches. The blade material was the aluminum alloy, which is usually selected for icing wind tunnel tests because of its good and stable thermal conductivity. According to the size of test section of wind tunnel, the test model has the chord length of 150 mm and the thickness or spanwise length of 20 mm. For this test model, the airfoil profile is consistent along spanwise, and its thickness is small. Therefore, it can be seen as a 2D blade airfoil in the present experimental research. The 3D effect of it can be negligible.

2.3 Test conditions

Based on the research object, several test parameters were selected for comparison. It included ambient temperatures, wind speeds, and icing times. To check the effect of an angle of attack (AOA) on the blade icing distribution, two kinds of angle

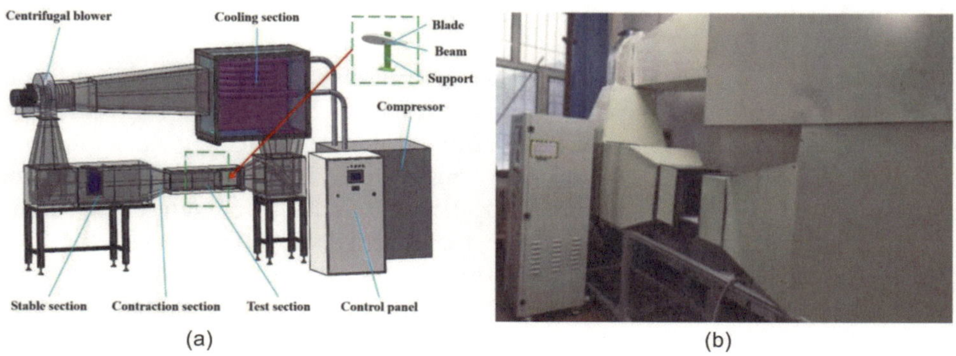

Figure 2.
Icing wind tunnel experimental system used in this study. (a) Schematic diagram of the icing wind tunnel experimental system. (b) Photo of the icing wind tunnel.

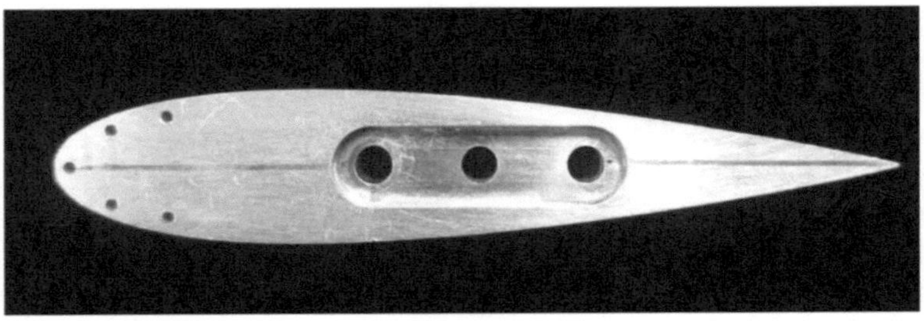

Figure 3.
Test model of the blade with NACA0018 airfoil.

Angel of attack (°)	Wind speed (m/s)	Temperature (°C)	Icing time (min)
0	5, 10	−5, −10, −15	0, 2, 4
10	5, 10	−5, −10, −15	0, 2, 4

Table 1.
Experimental conditions of the icing tests in the icing wind tunnel.

of attacks (AOAs), such as 0°and 10°, were selected. The test conditions are listed in **Table 1**. Furthermore, the LWC in the icing wind tunnel used in the present study was in the range of 0.8 ~ 1.6 g/m^3. The MVD was about 65 μm. In this study, the effects of LWC and MVD on icing of blade airfoil were not researched, which will be carried out deeply in the follow-up study. When the spray system begins to work, a high-speed camera (Phantom v5.1, with the revolution ratio of 1024 × 1024 pixels) begins to record the process of icing on blade surface and take photos at the moments of 2 minutes and the 4 minutes. By using the drawing software, the profile of icing shape can be obtained and analyzed.

3. Results and discussion

3.1 Icing distributions on the leading edge of blade surface

Figures 4 and **5** show the photos of icing distribution at the leading edge of blade under all test conditions. According to the figures, it can be seen that icing occurs on the leading edge of blade for all test conditions. It means that icing will appear when certain environmental conditions are met. However, the characteristics of icing distribution and ice accretion are obviously different under different test conditions.

For the effects of ambient temperature, the most typical characteristic is the change of icing type. When the temperature was −5°C, the type of icing on the blade surface was glaze ice with a transparent color. With the decrease in temperature, the type of rime ice, whose color was white, gradually began to appear. At the temperature of −10°C, there was a little rime ice, which only appeared on the upper surface away from the leading edge. This phenomenon was more obvious at the AOA of 0° than the one at 10°. When the temperature reached to −15°C, most of the ice transformed into the rime ice, while there was very little glaze ice, which only appeared at the front of the leading edge. The glaze ice was just formed by condensation of newly flowing supercooled water droplets in the ultimate period of photography. These

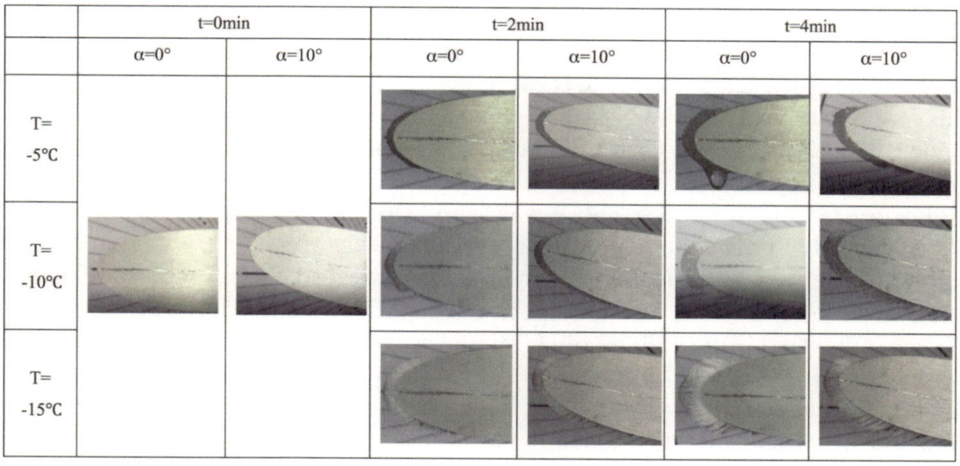

Figure 4.
Photos of icing distribution on blade at the wind speed of 5 m/s.

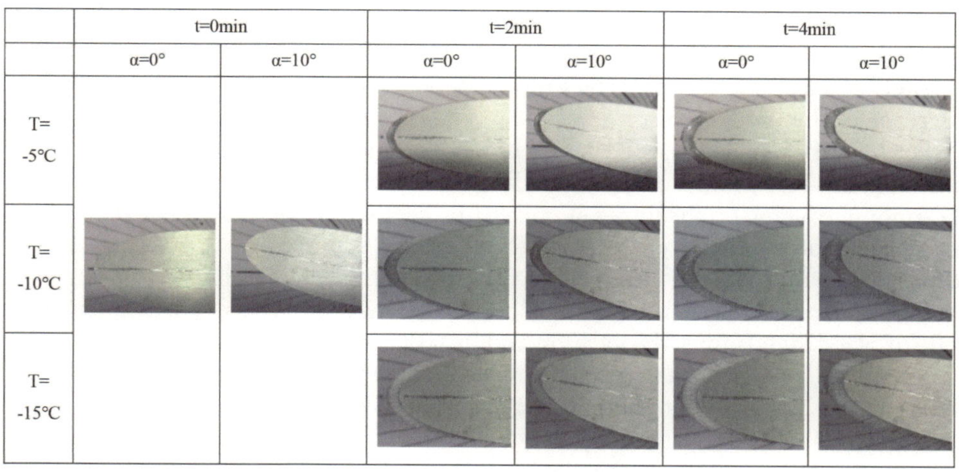

Figure 5.
Photos of icing distribution on blade at the wind speed of 10 m/s.

results indicate that the icing type on the leading edge of blade is glaze at high temperatures, such as above −5°C, and transforms into mixed ice and rime ice with the decrease in temperature. Furthermore, another phenomenon should be mentioned. For the AOA of blade at 0°, when the temperature was −5°C, there was a short icicle appearing on the lower surface near the leading edge of the blade at the icing time of 4 minutes. The reason for an explanation of this result is that the heat transfer rate of water droplets decreased with the increase in icing time. At the initial icing stage, the water droplets impinged directly on the blade surface whose temperature was the same as the environmental temperature. In this case, the water droplets froze in a short time because of higher heat conductivity coefficient of aluminum material (237 W/m·K) and low blade surface temperature. That is why the type of rime ice is generated on the blade surface. After that, there was a layer of ice covering the blade surface. The consequent water droplets impinged on the ice layer. The water droplets could not freeze in a short time because of low heat conductivity of ice (2.22 W/m·K).

Some super-cooled water droplets attaching on the ice layer surface had not enough time to freeze into ice completely due to the high temperature. These water droplets ran back along the blade surface to the trailing edge, and the icicle was generated. In this case, the type of icing was glaze ice due to high temperature of the icing blade. Under the action of wind speed and gravity, these incompletely frozen water droplets grow and form an ice icicle obliquely downward.

For the effects of wind speed, it was found that the wind speed does not play a key role in the change of icing type. When the temperature was constant, the type of icing was basically the same. When the wind speed increased from 5 m/s to 10 m/s, the icing amount decreases slightly. Meanwhile, the icing shape became smoother, especially at the temperature of −15°C. This may be caused by the constant flow rate of the spray nozzle in unit time. In the present study, the pressure of the pump is constant, which makes the flow rate of the nozzle constant in spite of increasing wind speed. Therefore, the water content passing through a cross-section is constant in unit time. In theory, the icing amounts are the same under two kinds of wind speeds. However, when the wind speed is higher, more water droplets with lightweight are blown away by high-speed wind, and the amount of water droplets impinging on the blade decreases. Additionally, the heat exchange of water droplets with air also accelerates and enhances because of the increase in wind speed. In this case, some small water droplets freeze before impinging on the blade surface, which also results in the decrease in the icing amount.

For the different AOAs, the amount of icing on the lower surface of blade increased because there was an upward angle of attack against wind. In contrast, the amount of icing on the upper surface had a decreasing trend. The key reason is that the windward state of the blade changes from symmetry to asymmetry due to the existence of AOA, which leads to the asymmetry of icing distribution on the leading edge of blade. Therefore, it was concluded that the AOA, or the windward state of blade, determines the location and distribution of icing on blade surface. Furthermore, a deep discussion can be made on how to use these results to the real wind turbine blade. As is known to all, the wind turbine blades always operate in a rotating state. Although the blade model in this research was in a static state, the research findings can be used for the analysis of rotating blades. Based upon the aerodynamics of wind turbine blade, the wind speed and the rotational speed will form a resultant speed at the local element of blade airfoil, which is known as relative wind speed. The AOA in the rotating condition should be decided by the relative wind speed. Therefore, the research findings obtained in the static condition can be used for the analysis of rotating blades.

3.2 Icing profile and icing area

In order to quantitatively analyze the distribution characteristics of icing, the captured icing photos were processed and the icing contours at the different moments were obtained, which are shown in **Figures 6** and 7.

For comparatively analyzing with the results at the AOA of 0 degrees, the figures of the icing blade at the AOA of 10° are turned back 10 degrees and located at horizontal level for convenient observation and comparison.

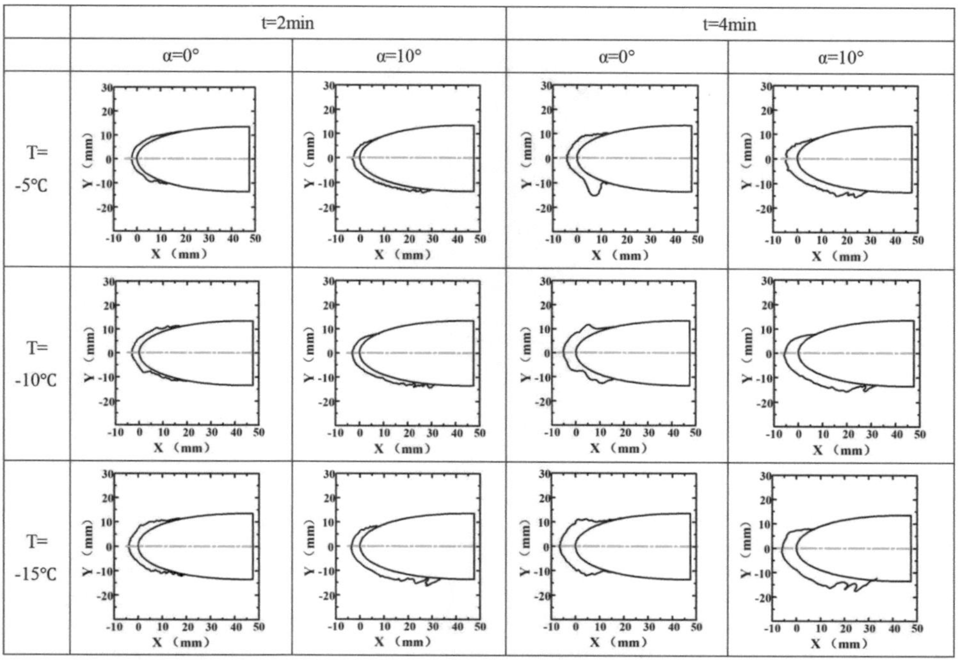

Figure 6.
Ice accretion on blade at the wind speed of 5 m/s.

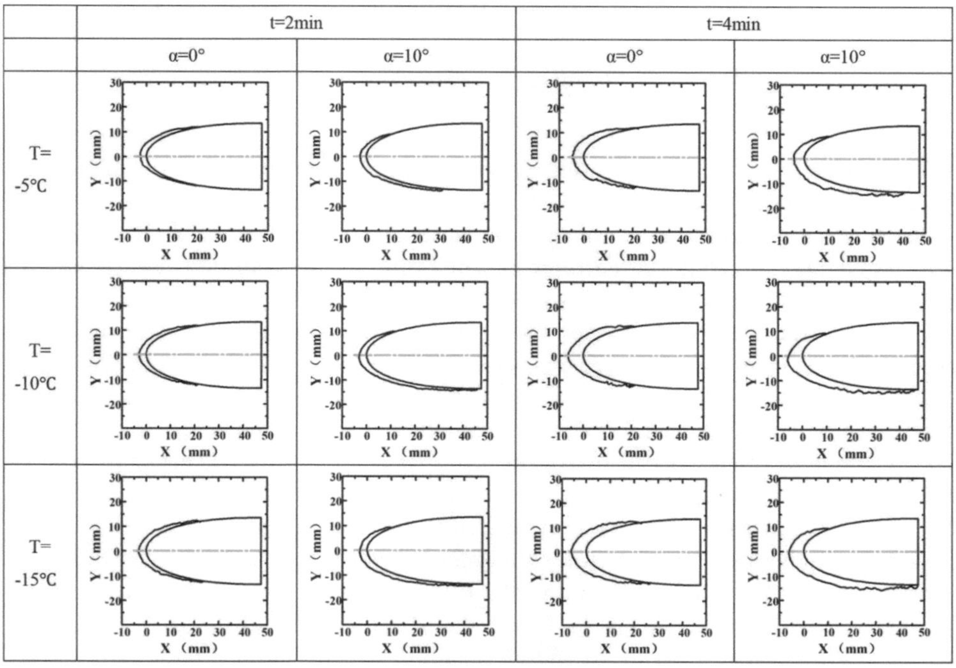

Figure 7.
Ice accretion on blade at the wind speed of 10 m/s.

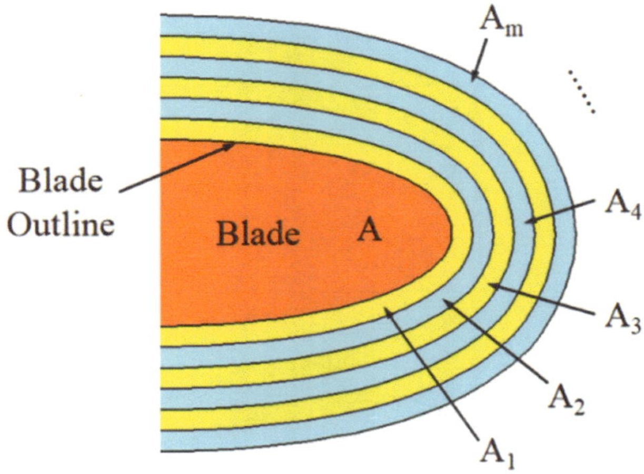

Figure 8.
Schematic diagram of icing area.

Based on these pictures showing the profile of icing, the change of icing shape can be seen clearly. Furthermore, from these figures, the icing area (A_i) is calculated in the present study. The schematic diagram of icing area is shown in **Figure 8**.

Figure 9 shows the changes in the icing area with the increase in icing time under all test conditions. Additionally, a dimensionless parameter, icing area ratio (η_s), can also be defined to quantitatively analyze the icing area. It is defined by the ratio of icing area (A_i) and the blade airfoil area (A), which is expressed in Eq. (1).

$$\eta_s = \frac{A_i}{A} \times 100\% \tag{1}$$

The variations of icing area ratios with icing time, calculated from the data in **Figure 9**, are shown in **Figure 10**.

As shown in **Figure 9**, the icing area increased linearly with the increase in icing time in general. This characteristic can also be found in the result of the icing area ratio in **Figure 10**. Therefore, in this section, it only focuses on the analysis of icing area ratio in **Figure 10**. In the period of 4 minutes icing time, for all conditions in this research, the maximum icing area ratios reached to 6.79% and 8.31% at the AOAs of 0° and 10°, respectively. At the wind speed of 5 m/s, with the decrease in temperature, the growth rate of icing area ratio increased. This phenomenon was obvious, especially at the AOA of 10°. However, for the wind speed of 10 m/s, the effect of temperature on growth rate of icing area ratio became more obvious than that at a wind speed of 5 m/s. The reasons for this result are multifaceted and complex. The high wind speed gives high kinetic energy to the supercooled water droplets, which makes the impact course of water droplets on the blade surface complicated. Also, higher wind speed will affect the characteristics of heat transfer more obviously in comparison with that of the lower wind speed. When the temperature is low, a higher wind speed intensifies the heat transfer of water droplets, which enhances the growth rate of icing area. Deep researches on this issue will be carried out in our further tests.

More discussions were given on the effects of icing time and angle of attack. For the icing time, the icing time length of 4 minutes is selected in this study. Of course, with the increase in icing time, the icing area will increase. In nature, the process of

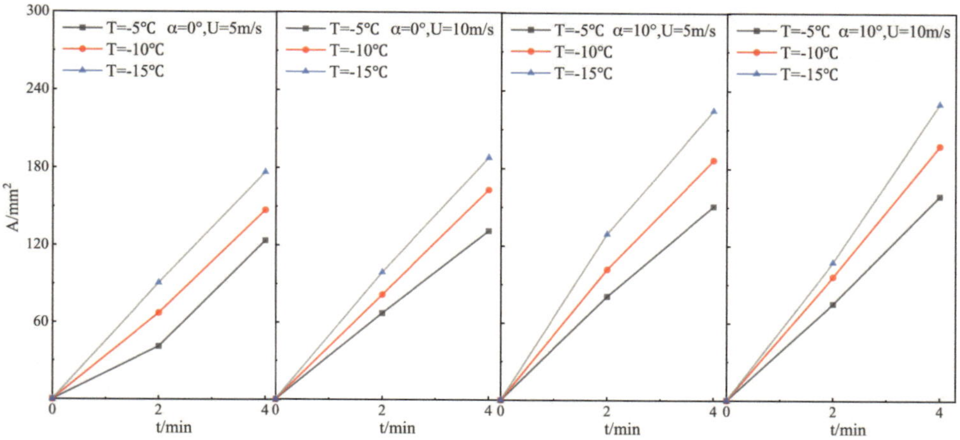

Figure 9.
Variation of icing area with icing time.

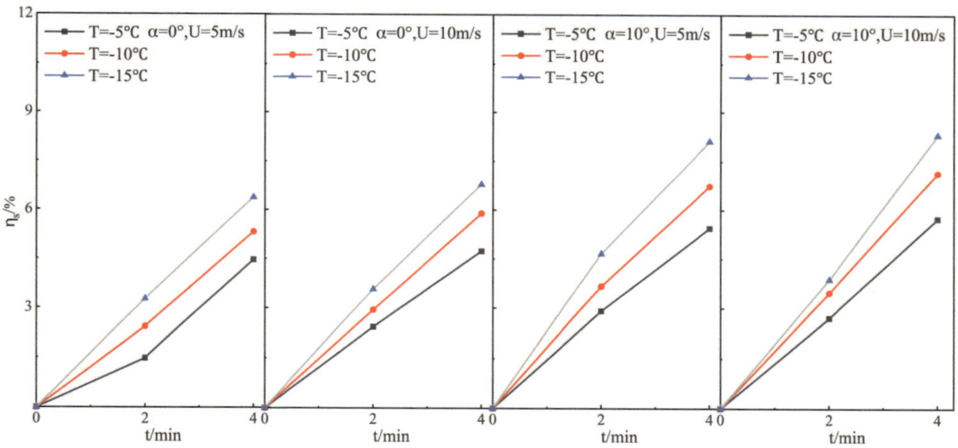

Figure 10.
Variation of icing area ratio with icing time.

icing on a wind turbine is a long course, which takes time from hours to days. In this research, the main purpose is to explore the initial stage of icing on the leading edge of blade surface. When the ice forms to a certain shape and thickness, the consequent icing process occurs between supercooled water droplets and ice formed on the blade surface, not the blade surface. For the angle of attack, the most important effect is to change the windward area of the blade. Based on the test results of AOA, it is concluded that the icing area will increase along with AOA. For the blade in a static condition, the range of AOA is wide. However, for the rotating blade of a real wind turbine, the relative angle of attack is controlled in a limited range. Therefore, the test results can be a useful reference to the icing research of a real wind turbine.

4. Conclusions

In this research, the icing tests of a blade model for a wind turbine were carried out in an icing wind tunnel. The distribution characteristics of icing on the leading edge of the blade were obtained and analyzed. Some key results in the present study are summarized and listed as follows:

1. In the present study, the icing on the leading edge of a blade with NACA0018 airfoil was explored and obtained. It is concluded that the ice type has good relationship with the ambient temperature. With the decrease in temperature, the ice type changes from glaze ice, mixed ice, to rime ice. Meanwhile, the phenomenon is more obvious when the wind speed increases. Additionally, with the increase in wind speed, the icing amount of the blade decreases slightly as the flow rate of spray nozzle is constant. Under low-temperature condition, high wind speed can accelerate the growth rate of icing.

2. When the environmental condition keeps constant, the icing area increases linearly along with the icing time. Additionally, the icing area is also affected by the angle of attack. With the increase in angle of attack, the windward area of the blade increases, which results in an increase in the icing area. Moreover, the area covered by ice on the lower surface increases with respect to the AOA. The maximum icing area ratios reach to 6.79% and 8.31% at the AOAs of 5° and 10°, respectively.

Based on the method of icing wind tunnel test in this research, the characteristics of icing on blade surface can be researched and obtained. This study can be a reference to the icing mechanism research and the anti- and de-icing research of wind turbines.

Acknowledgements

This work is supported by the National Natural Science Foundation of China (NSFC) [grant number 51976029]. The authors would like to thanks for the support.

Author details

Yan Li[1*], Zhongqiu Mu[1], Zhiyuan Liu[1], Wenfeng Guo[1], Fang Feng[1] and Kotaro Tagawa[2]

1 Northeast Agricultural University, Harbin, China

2 Tottori University, Japan

*Address all correspondence to: liyanneau@163.com

References

[1] Amjith LR, Bavanish B. A review on biomass and wind as renewable energy for sustainable environment. Chemosphere. 2022;**293**:133579

[2] Li Y. Straight-bladed vertical Axis wind turbines: History, performance, and applications. Rotating Machinery. 2020:87-103

[3] Li Y, Zhao S, Chunming Q, Tonga G, Feng F, Zhao B. Aerodynamic characteristics of straight-bladed vertical Axis wind turbine with a curved-outline wind gathering device. Energy Conversion and Management. 2020;**203**:112249

[4] Dalili N, Edrisy A, Carriveau R. A review of surface engineering issues critical to wind turbine performance. Renewable and Sustainable Energy Reviews. 2009;**13**(2):428-438

[5] Li Y, Wang S, Liu Q, Feng F, Tagawa K. Characteristics of ice accretions on blade of the straight-bladed vertical axis wind turbine rotating at low tip speed ratio. Cold Regions Science and Technology. 2018;**145**:1-13

[6] Wenfeng G, He S, Yan L. Wind tunnel tests of the rime icing characteristics of a straight-bladed vertical axis wind turbine. Renewable Energy. 2021;**179**:116-132

[7] Alessandro Z, Michele DG, Helmut K. Wind energy harnessing of the NREL 5 MW reference wind turbine in icing conditions under different operational strategies. Renewable Energy. 2018;**115**:760-772

[8] Ibrahim GM, Pope K, Muzychka YS. Effects of blade design on ice accretion for horizontal axis wind turbines. Journal of Wind Engineering and Industrial Aerodynamics. 2017;**173**:39-52

[9] Li Y, Shi L, Guo W, Tagawa K, Zhao B. Numerical simulation of icing effect on aerodynamic characteristics of a wind turbine blade. Thermal Science. 2021;**25**(6B):4643-4650

[10] Yan Lie Ce Sun, Yu Jiang, Fang Feng. Scaling method of the rotating blade of a wind turbine for a rime ice wind tunnel test. Energies. 2019;**12**:626-627

[11] Xian Y, Kai-chun W, Hong-lin M, Guo-lin Z. 3-D numerical simulation of droplet collection efficiency in large-scale wind turbine icing. Acta Aerodynamica Sinica. 2013;**31**(06): 745-775 (In Chinese)

[12] Yan L, He S, Wenfeng G. Effect of ultrasonic vibration on the surface adhesive characteristic of iced Aluminum alloy plate. Applied Sciences-Basel. 2022;**12**(5):2357

[13] Yan L, Dong X, Guo W, etc. Ultrasonic micro-vibration deicing of flat plate for wind turbine blades. Journal of Drainage and Irrigation Machinery Engineering. 2022;**40**(2):204-210 (In Chinese)

[14] Li X-j, Guo W-f, Li Y, Zhi X, Feng F. Wind tunnel test of an anti-icing approach by heat pipe for wind turbine blades under the rime ice condition. Thermal Science. 2021;**25**(6B):4485-4493

[15] Zeng J, Song B. Research on experiment and numerical simulation of ultrasonic de-icing for wind turbine blades. Renewable Energy. 2017;**113**:706-712

[16] Xian Y, Yewei G, Guolin Z, et al. Experimental and computational

investigation into ice accretion on airfoil of a transport aircraft. Journal of Aerospace Power. 2011;**26**:808-813 (In Chinese)

[17] Yan L, Long WS, Ce S. Icing distribution of rotating blade of horizontal axis wind turbine based on Quasi-3-D numerical simulation. Thermal Science. 2018;**22**:S681-S691

[18] Yan L, Ce S, Jiang Y, Xian Y, Wenfeng G, Shaolong W, et al. Influence of liquid water content on wind turbine blade icing by numerical simulation. Journal of Drainage and Irrigation Machinery Engineering. 2019;**37**(6):513-520

[19] Li Y, Shen H, Guo W. Simulation and experimental study on the ultrasonic micro-vibration De-icing method for wind turbine blades. Energies. 2021;**14**(24):8246

[20] Guo W, Shen H, Li Y, Feng F, Tagawa K. Wind tunnel tests of the rime icing characteristics of a straight-bladed vertical axis wind turbine. Renewable Energy. 2021;**179**:116-132

[21] Lei S, Fang F, Wenfeng G, Yan L. Research and Development of a small-scale icing wind tunnel test system for blade Airfoil icing characteristics. International Journal of Rotating Machinery. 2021;**2021**:5598859

Chapter 4

A Review of Signal Analysis Methods and Their Applications in the Reversible Pump Turbine

Hao Li, Xianghao Zheng, Yuning Zhang, Jinwei Li and Yuning Zhang

Abstract

This chapter summarizes the signal analysis methods and their applications in the field of reversible pump turbine, including time-domain analysis, time-frequency analysis, mode decomposition and de-noising. The time-domain analysis methods include the time-domain statistical parameters, correlation analysis and chaos evaluation. The time-frequency analysis methods include the short-time Fourier transform, Hilbert-Huang transform and VMD-Hilbert transform. Furthermore, the signal mode decomposition and signal de-noising will be introduced together with the related evaluation indexes. The applications of the aforementioned methods are demonstrated based on both the simulated ideal signals and the measured signals from the prototype reversible pump turbines.

Keywords: time-domain analysis, time-frequency analysis, mode decomposition, signal de-noising, reversible pump turbine

1. Introduction

There are a large number of rotating machineries being intensively employed in the modern industry, such as hydro-turbines, pumps, steam turbines, compressors, and generators. They are widely employed in the power generation, heating, refrigeration, and chemical industries. The rotating machineries often involve extreme conditions such as extremely low load, high temperature, high pressure, high speed, and overload. Hence, the malfunctions may occur after the long-term operation with possibly catastrophic consequences. Therefore, it is very important to understand and monitor the operational states of rotating machineries.

With the developments of data analysis technology, signal analysis methods become gradually mature and have been widely applied in the field of condition monitoring and fault diagnosis of various kinds of rotating machineries. The signal analysis could extract useful information from the original signal, and judge the operational states of the equipment based on the obtained information. Signal analysis is of great significance to the rotating machineries both for condition monitoring and fault diagnosis.

A reversible pump turbine is the core part of the pumped hydro energy storage power station. It can switch between the pumping and the generating mode according to the actual demand. When the reversible pump turbine deviates from the design working condition, it is easy to produce serious pressure pulsation and vibration due to the influence of rotor-stator interaction in the vaneless space and the vortex rope in the draft tube, which will affect the normal operation of the pump turbine and the safety of the whole power station [1–3]. Therefore, to monitor the operational states of reversible pump turbine and ensure its safety and efficient operation, it is very necessary to employ the signal analysis methods extensively.

This chapter will review several typical signal analysis methods and introduce their applications in the field of the reversible pump turbine in detail through specific cases. This chapter will be divided into the following four parts. The first part will introduce the time-domain analysis methods of the signals. The second part will introduce the time-frequency analysis methods of the signals. The third part will introduce the signal decomposition and the signal de-noising. The fourth part will demonstrate the applications of the introduced signal analysis methods in the field of the reversible pump turbines with the aid of on-site measured signals.

2. Time-domain analysis methods

Signals can convey information by expressing the relationship between time and other physical quantities. The time-domain analysis methods of signals refer to a series of processing such as amplification, filtering, statistical feature calculation and correlation analysis in the time domain. Through time-domain analysis methods, the characteristic parameters reflecting the operational state of the mechanical equipment can be extracted from the signal, which could be further employed for the purpose of the evaluation of the operational state and fault diagnosis of the equipment. Time-domain analysis methods could also analyze the auto-correlation (AC) and cross-correlation characteristics of the signal together with the degree of chaos. This section will be divided into three parts including the time domain statistical parameters, the correlation analysis and the chaos evaluation.

2.1 Time-domain statistical parameters

The classical time-domain statistical parameters include the peak value, peak-to-peak value, mean value, mean square value, root mean square, variance and standard deviation. The meanings of these statistical parameters are explained in detail below. Here, the $x(t)$ is the signal varying with the time. The t_1 and t_2 are the starting and the ending time of the signal in the time domain.

Peak value

Peak value refers to the maximum amplitude of a signal.

$$x_p = \max|x(t)| \tag{1}$$

Peak-to-peak value

Peak-to-peak value refers to the difference between the maximum value and the minimum value of a signal within a given period. In the practical application, to eliminate the influences of distortion and noise in the signal, the peak-to-peak value is usually calculated by the confidence interval method with employing the confidence coefficient (e.g., 97 or 95%). Then, the peak-to-peak value could be calculated by using the upper and the lower limits of the obtained signal.

Mean value

The mean value is the averaged value of the whole signal.

$$\overline{x} = \frac{1}{t_2 - t_1} \int_{t_1}^{t_2} x(t)dt \tag{2}$$

Mean square value

Mean square value refers to the mean of the square of the signal, indicating the strength and average power of the signal.

$$x_{ms} = \frac{1}{t_2 - t_1} \int_{t_1}^{t_2} x^2(t)dt \tag{3}$$

Root mean square

Root mean square refers to the arithmetic square root of the mean square value.

$$x_{rms} = \sqrt{\frac{1}{t_2 - t_1} \int_{t_1}^{t_2} x^2(t)dt} \tag{4}$$

Variance

Variance is the average square value of the difference between the amplitude of the original signal and the mean value. It describes the fluctuation range of the signal and represents the strength of the fluctuating component in the signal.

$$s^2 = \frac{1}{t_2 - t_1} \int_{t_1}^{t_2} (x - \overline{x})^2 dt \tag{5}$$

Standard deviation

Standard deviation is the arithmetic square root of variance.

$$s = \sqrt{\frac{1}{t_2 - t_1} \int_{t_1}^{t_2} (x - \overline{x})^2 dt} \tag{6}$$

2.2 Correlation analysis

When there are multiple signals, sometimes it is necessary to study the relationship between them. Correlation analysis (including auto-correlation function and cross-correlation analysis) could show the similarity or dependence between signals.

Auto-correlation function

The auto-correlation (AC) function describes the correlation of a signal between its values at a certain time and after a certain time delay τ. It can quantitatively describe the degree of similarity between the time-domain waveform of a signal after shifting τ on the time and the original time-domain waveform of a signal.

$$R_{xx}(\tau) = \int_{-\infty}^{\infty} x(t)x(t + \tau)dt = \int_{-\infty}^{\infty} x(t - \tau)x(t)dt \tag{7}$$

$R_{xx}(\tau)$ is an even function. When $\tau = 0$, $R_{xx}(\tau)$ reaches its maximum value and is equal to the energy of the signal.

Cross-correlation function

The cross-correlation function describes the degree of similarity of the time-domain waveforms of two signals $x(t)$ and $y(t)$ at t and $t + \tau$.

$$R_{xy}(\tau) = \int_{-\infty}^{\infty} x(t)y(t+\tau)dt = \int_{-\infty}^{\infty} x(t-\tau)y(t)dt \tag{8}$$

The higher the $R_{xy}(t)$ is, the higher the similarity between $x(t)$ and $y(t+\tau)$ is. It should be noted that $R_{xy}(\tau)$ is not equal to $R_{yx}(\tau)$, and the relationship between them is as follows:

$$R_{xy}(\tau) = R_{yx}(-\tau) \tag{9}$$

Normalization of the correlation function

In the practical application of signal correlation analysis, to compare the magnitude of the correlation, we often use the normalized form of the correlation function. The normalized form of the auto-correlation function is defined as follows:

$$\rho_{xx}(\tau) = \frac{R_{xx}(\tau)}{\sqrt{R_{xx}(0)}\sqrt{R_{xx}(0)}} = \frac{R_{xx}(\tau)}{R_{xx}(0)} \tag{10}$$

The normalized form of the cross-correlation function is defined as follows:

$$\rho_{xy}(\tau) = \frac{R_{xy}(\tau)}{\sqrt{R_{xx}(0)}\sqrt{R_{yy}(0)}} \tag{11}$$

where $|\rho_{xx}(\tau)| \leq 1, \left|\rho_{xy}(\tau)\right| \leq 1$. When $\rho(\tau) = 1$, it means that the correlation between the two signals reaches the maximum; when $\rho(\tau) = 0$, it means that two signals are not correlated; when $\rho(\tau) = -1$, it means that two signals are linearly negatively correlated.

2.3 Measurement of chaos

Chaos is an inherent characteristic of a nonlinear dynamic system. When a signal is generated by a nonlinear system, it may show the characteristic of chaos. At this time, the signal waveform is very irregular and is very similar to the characteristic of random noise. Entropy can be adopted to measure the chaos of a nonlinear dynamic system. The more chaotic the system, the higher its entropy. Therefore, we can use entropy to evaluate the degree of chaos in a signal. As a typical method, the permutation entropy for evaluating the degree of chaos in a signal is introduced below [4].

Permutation entropy

Permutation entropy is an index to measure the complexity of the time series [4]. The more regular the time series is, the smaller the permutation entropy is. The corresponding calculation process is given as follows [4].

Considering a time series $x = \{x_i, i = 1, 2, \ldots, N\}$, it can be reconstructed into the following form [4]:

$$\begin{cases} X_1 = \{x_1, x_{1+\tau}, \dots, x_{1+(em-1)\tau}\} \\ \vdots \\ X_i = \{x_i, x_{i+\tau}, \dots, x_{i+(em-1)\tau}\} \\ \vdots \\ X_{N-(em-1)\tau} = \{x_{N-(em-1)\tau}, x_{N-(em-2)\tau}, \dots, x_N\} \end{cases} \tag{12}$$

Here, the "N" is the total number of the data points within the whole signal. The "τ" refers to the delay time and the "em" refers to the embedded dimension.

Now, each X_i could be sorted in the ascending order internally [4]:

$$x_{i+(j_1-1)\tau} \leq x_{i+(j_2-1)\tau} \leq \dots \leq x_{i+(j_{em}-1)\tau} \tag{13}$$

Here, the "$j_1, j_2, \dots, j_{em}$" in the subscripts refer to the indexes for different columns. Thus, a sequence of symbols can be obtained for any vector X_i [4]:

$$S(h) = [j_1, j_2, \dots, j_{em}] \tag{14}$$

There are up to "$em!$" kinds of such index sequences. The occurrence probability of each index sequence is calculated and expressed as $P_1, P_2, \dots, P_r$, respectively. Here, $h = 1, 2, \dots, r$ and $r \leq em!$. Then, the permutation entropy of the original signal can be expressed as follows [4]:

$$H_p(em) = -\sum_{h=1}^{r} P_h \ln P_h \tag{15}$$

When $P_h = 1/em!$, the above formula can reach the maximum value $\ln(em!)$. Therefore, the normalized form of Eq. (15) can be obtained as follows [4]:

$$H(em) = \frac{H_P(em)}{\ln(em!)} \tag{16}$$

The variation range of the normalized permutation entropy is between 0 and 1. In the practical application of calculating the permutation entropy of a signal, the whole data points within the signal can be intercepted by a window with a certain length. And the permutation entropy of the data points in each window is calculated by moving the window one data at a time. In the present chapter, the average permutation entropy of all the windows will be taken as the permutation entropy of the whole signal.

2.4 Examples of a signal time-domain analysis

In this section, the pressure pulsation signal (the pressure pulsation signal 1) in the vaneless space of a prototype pump turbine under the dimensionless load condition (the ratio of the actual operational load to the rated load) P^*=52% is taken in the analysis to demonstrate the applications of the time-domain analysis method introduced in Sections 2.1–2.3. The length of the signal is 1 s and the sampling frequency is 2000 Hz. In this section, p^* refers to the dimensionless pressure pulsation using the water head (326.01 m). **Figure 1** shows the dimensionless pressure pulsation 1 in the time domain.

Time-domain statistical parameters

The statistical parameters of the dimensionless pressure pulsation signal 1 in the time domain are shown in the **Table 1** as follows:

Auto-correlation analysis

Figures 2 and **3** show the normalized AC functions of the dimensionless pressure pulsation signal 1 and a random noise signal. When $\tau = 0$, the AC function of the dimensionless pressure pulsation signal 1 in the **Figure 2** reaches its maximum value, but it gradually decreases to near 0 on both sides of $\tau = 0$. For the random noise signal, its AC function reaches the maximum also at $\tau = 0$. When $\tau \neq 0$, due to the

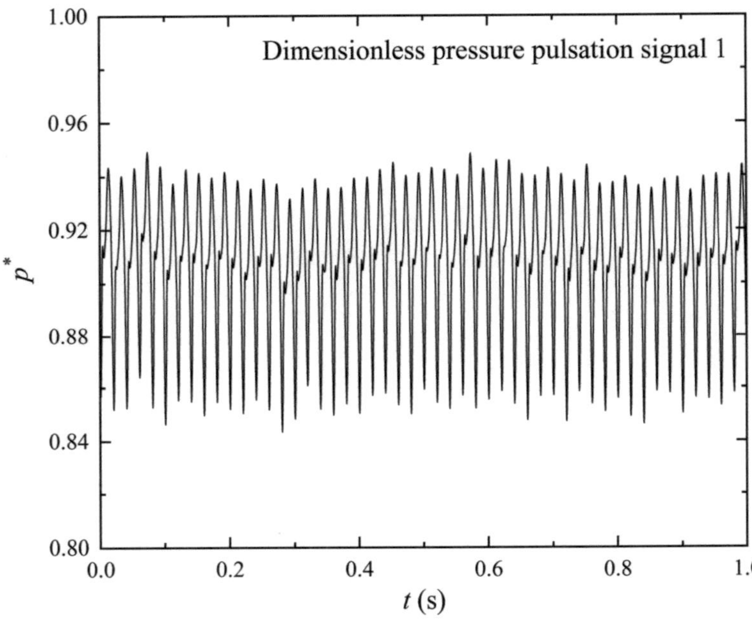

Figure 1.
The time-domain diagram of the dimensionless pressure pulsation signal 1.

Statistical parameter	Value
Peak value	0.9492
Peak-to-peak value (confidence coefficient 97%)	0.0911
Peak-to-peak value (confidence coefficient 95%)	0.0883
Mean	0.9081
Mean square value	0.8252
Root mean square	0.9084
Variance	0.0006
Standard deviation	0.0244

Table 1.
The time-domain statistical parameters of the dimensionless pressure pulsation signal 1.

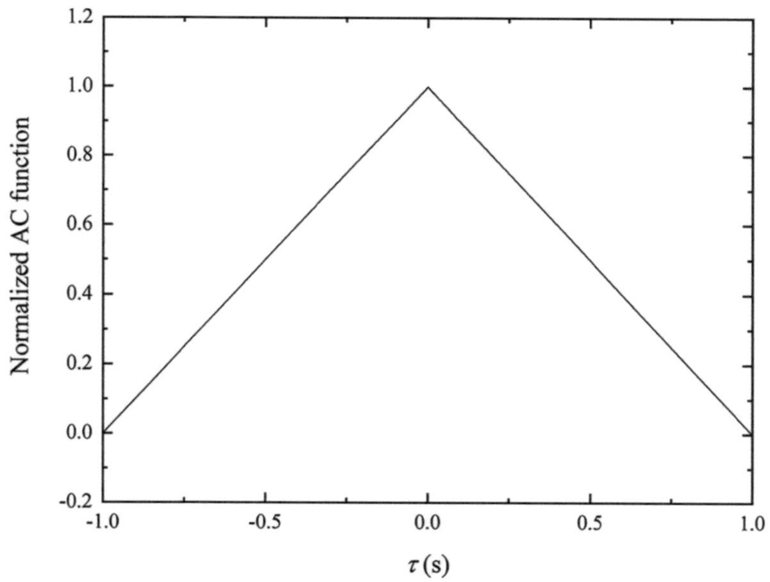

Figure 2.
The normalized AC function of the dimensionless pressure pulsation signal 1.

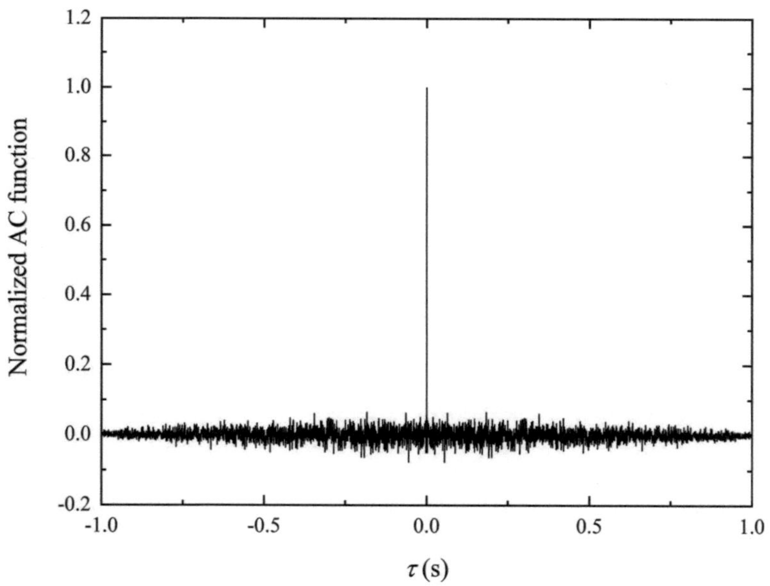

Figure 3.
The normalized AC function of the random noise.

randomness of noise, its AC function decreases rapidly. Therefore, the normalized AC function of the random noise in **Figure 3** is close to 1 at $\tau = 0$ and approximately equal to 0 on both sides of $\tau = 0$.

Cross-correlation analysis

In the experiment, another dimensionless pressure pulsation signal (the dimensionless pressure pulsation signal 2) was measured in the vaneless space at the same

Signal	Permutation entropy
Dimensionless pressure pulsation signal 1	0.4254
Dimensionless pressure pulsation signal 2	0.4271
Random noise	0.5666

Table 2.
Analysis results of permutation entropy of different types of signals.

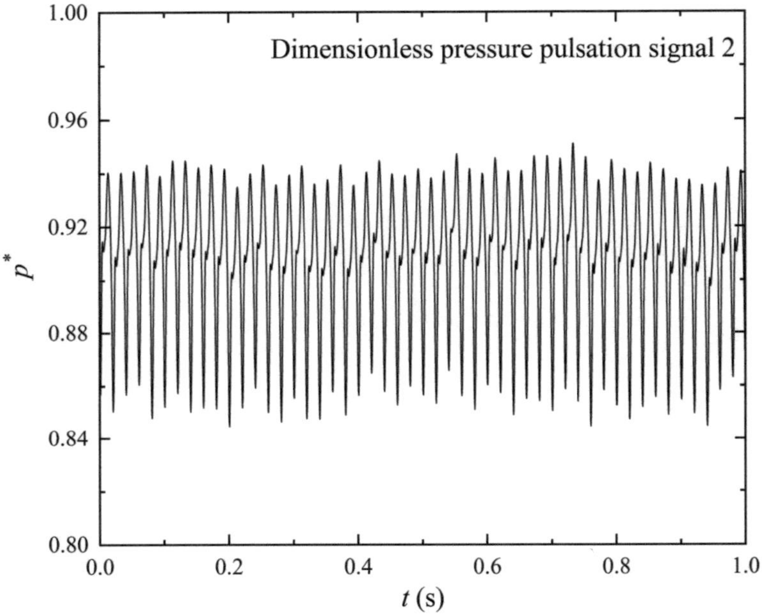

Figure 4.
The time-domain diagram of the dimensionless pressure pulsation signal 2.

operational condition (P^*=52%) during the different time period. **Figure 4** shows the time-domain diagram of the dimensionless pressure pulsation signal 2.

When the delay time $\tau = 0$, the normalized cross-correlation function of the dimensionless pressure pulsation signals 1 and 2 is 0.98, indicating that there is a strong correlation between them. On the contrary, when the delay time $\tau = 0$, the normalized cross-correlation function of the dimensionless pressure pulsation signal 1 and random noise signal is 0.0088, indicating that there is almost no correlation between them.

Permutation entropy analysis

In the present chapter, the embedded dimension, delay time and window length required in the analysis procedure of the permutation entropy are 6, 3 and 128, respectively. The analysis results of the permutation entropy of the dimensionless pressure pulsation signal 1, the dimensionless pressure pulsation signal 2 and the random noise signal are as follows:

It can be seen from the **Table 2** that the analysis results of the permutation entropy of the dimensionless pressure pulsation signals 1 and 2 are smaller than that of the random noise, because the dimensionless pressure pulsation signals contain certain characteristic frequencies with regularity. The permutation entropy of the random noise is large, which indicates its high degree of chaos.

3. Time-frequency analysis methods

The aim of the time-frequency analysis is to study the time-varying signals, which can reflect the relationship of the variations of the frequency and amplitude with time within the signal. Generally speaking, the typical time-frequency analysis methods include the short-time Fourier transform [5], the Hilbert-Huang transform [6] and the VMD-Hilbert transform with a detailed introduction in this section.

3.1 The traditional Fourier transform

Before introducing the time-frequency analysis methods, the traditional Fourier transform is:

$$F(\omega) = \int_{-\infty}^{\infty} x(t)e^{-j\omega t}dt \tag{17}$$

where $x(t)$ is the original signal and $e^{-j\omega t}$ is the basis function of the Fourier transform. The traditional Fourier transform can accurately reflect the frequency components contained in the signal, but it cannot reflect the time information of these frequency components.

3.2 The time-frequency analysis methods

Short-time Fourier transform

$$\mathrm{STFT}_x(\tau, \omega) = \int_{-\infty}^{\infty} x(t)g(\tau - t)e^{-j\omega t}dt \tag{18}$$

The formula of the short-time Fourier Transform (STFT) is presented above [5]. Compared with the traditional Fourier transform, the window function is added to the formula. STFT needs to select a window function, which can be moved continuously to identify the frequencies at different times. The window function will affect the resolution of the spectrum. The smaller the window size, the higher the time resolution of the spectrum. Normally, the low frequency band requires high frequency resolution and low time resolution. And, the high frequency band requires low frequency resolution and high time resolution. However, after the window function is selected, the resolution of the entire spectrum is fixed and cannot be adjusted. Therefore, STFT is suitable for analyzing the segmented stationary signals or approximately stationary signals.

Hilbert-Huang transform

Hilbert-Huang transform (HHT) consists of two parts including the empirical mode decomposition (EMD) and the Hilbert spectrum analysis (HSA). These two parts will be introduced in detail below.

The EMD method can decompose the original signal into several intrinsic mode functions (IMFs) and the residual component $r(t)$. The specific steps are given as follow.

Firstly, calculating all the extreme points of the signal $x(t)$, and using the cubic spline function to interpolate and fit the maximum and minimum points respectively to obtain the upper envelope $a(t)$ and lower envelopes $b(t)$ of the signal;

Secondly, calculating the mean curve $c(t)$ of the upper and lower envelopes and the difference $h(t)$ between the signal $x(t)$ and $c(t)$ [6]:

$$c(t) = \frac{a(t) + b(t)}{2} \tag{19}$$

$$h(t) = x(t) - c(t) \tag{20}$$

Thirdly, the screening stop condition is given as [6]:

$$\sum_{t=0}^{T} \frac{\left|h_{g-1}(t) - h_g(t)\right|^2}{h_{g-1}^2(t)} \leq \varepsilon \tag{21}$$

Here, the "g" refers to the number of screenings. The "T" refers to the signal time length and the "ε" refers to a preset value. If $h(t)$ does not meet the condition, replacing$x(t)$ with $h(t)$ and repeating the above steps. And if $h(t)$ meets the condition, then $h(t)$ will be the first IMF to be extracted.

Fourthly, repeating the above steps to obtain all the remaining IMF until the residual component $r(t)$ becomes a constant value sequence with a frequency approximately 0.

Although, the EMD can realize signal decomposition, it has the problems of end-point effect and mode mixing. Endpoint effect means that analysis results at the signal endpoints will produce errors, which will affect the internal data. Mode mixing means that the EMD method cannot effectively separate different mode components based on the time characteristic scales. These problems can lead to serious performance degradation of EMD methods.

HSA performs a Hilbert transform on each IMF obtained by EMD. The specific steps are given as follows:

Firstly, performing the Hilbert transform on each IMF $h_i(t)$ [6]:

$$H[h_i(t)] = h_i(t) * \frac{1}{\pi t} = \frac{1}{\pi} \int_{-\infty}^{\infty} \frac{h_i(\tau)}{t - \tau} d\tau \tag{22}$$

Here, i represents the i-th IMF component; "*" represents the convolution operation.

Secondly, using the above formula to construct the analytical signal $d_i(t)$ of each IMF $h_i(t)$ [6]:

$$d_i(t) = h_i(t) + jH[h_i(t)] = a_i(t) e^{j\varphi_i(t)} \tag{23}$$

Here [6],

$$a_i(t) = \sqrt{h_i^2(t) + H^2[h_i(t)]} \tag{24}$$

$$\varphi_i(t) = \arctan\left\{\frac{H[h_i(t)]}{h_i(t)}\right\} \tag{25}$$

Thirdly, the instantaneous frequency of each IMF can be obtained from the following formula [6]:

$$w_i(t) = \frac{d\varphi_i(t)}{dt} \tag{26}$$

After expressing the results on the time-frequency plane, the Hilbert amplitude spectrum of each IMF $h_i(t)$ can be obtained as follows [6]:

$$H_i(w,t) = \begin{cases} 0, w \neq w_i(t) \\ a_i(t), w = w_i(t) \end{cases} \tag{27}$$

Fourthly, by presenting the amplitude spectrum of all IMFs in a spectrum, the Hilbert spectrum of the original signal can be obtained as follows [6]:

$$H(w,t) = \text{Re} \sum_{i=1}^{n} a_i(t) e^{j \int w_i(t) dt} \tag{28}$$

The advantage of HHT is that it is not restricted by the linear and stationary characteristics of the signal and can analyze non-linear and non-stationary signals. At the same time, the HHT is self-adaptive through using the EMD to eliminate the choice of basis function. In addition, HHT is not restricted by Heisenberg's uncertainty principle.

VMD-Hilbert transform

The principle of VMD-Hilbert transform is similar to HHT. It also uses the signal decomposition method to obtain the IMF component of the signal, and then obtains the Hilbert spectrum of the signal through HAS. The difference between these two is that the variational mode decomposition (VMD) [7] is adopted as the signal decomposition method.

The core idea of the VMD method is to construct and solve the variational optimization problem in the frequency domain. The specific steps include the following ones.

Firstly, decomposing the original signal into K IMFs with respective central angular frequencies and limited bandwidths [7]:

$$\text{IMF}_k(t) = A_k(t) \cos\left[\varphi_k(t)\right] \tag{29}$$

Here, $\varphi_k(t)$ refers to phase of the signal and $A_k(t)$ refers to instantaneous amplitude.

Secondly, the constraints are added to ensure that the sum of the estimated bandwidths of each IMF is minimum and the sum of K IMFs is equal to the original signal [7].

$$\begin{cases} \min\limits_{\{\text{IMF}_k\},\{\omega_k\}} \left\{ \sum\limits_{k=1}^{K} \left\| \partial_t \left[\left(\delta(t) + \frac{j}{\pi t} \right) * \text{IMF}_k(t) \right] e^{-j\omega_k t} \right\|_2^2 \right\} \\ s.t. \sum\limits_{k=1}^{K} \text{IMF}_k(t) = x(t) \end{cases} \tag{30}$$

Here, $\delta(t)$ refers to the Dirac function. ω_k is the central angle frequency of the k-th IMF_k. "*" refers to the convolution operation.

Thirdly, the Lagrange multiplication operator is introduced to transform the constrained variational problem into an unconstrained variational problem. And an augmented Lagrange expression is obtained as follows [7]:

$$L(\{\mathrm{IMF}_k\}, \{\omega_k\}, \lambda) = \alpha \sum_{k=1}^{K} \left\| \partial_t \left[\left(\delta(t) + \frac{j}{\pi t} \right) * \mathrm{IMF}_k(t) \right] e^{-j\omega_k t} \right\|_2^2 + \left\| \mathrm{IMF}(t) - \sum_{k=1}^{K} \mathrm{IMF}_k(t) \right\|_2^2 + \left\langle \lambda(t), \mathrm{IMF}(t) - \sum_{k=1}^{K} \mathrm{IMF}_k(t) \right\rangle \tag{31}$$

Here, α refers to the quadratic penalty factor.

Fourthly, the alternating direction multiplier method is used to iteratively solve IMF_k, ω_k and λ to find the saddle point of the unconstrained variational problem. The update formulas of $\hat{\mathrm{IMF}}_{k_{n+1}}(\omega)$, ω_k^{n+1} and $\hat{\lambda}^{n+1}(\omega)$ obtained by Parseval Fourier equidistant transformation could be expressed as [7]:

$$\hat{\mathrm{IMF}}_{k_{n+1}}(\omega) = \frac{\hat{x}(\omega) - \sum_{i=1}^{k-1} \hat{\mathrm{IMF}}_i^{n+1}(\omega) - \sum_{i=k+1}^{K} \hat{\mathrm{IMF}}_i^{n}(\omega) + \frac{\hat{\lambda}^n(\omega)}{2}}{1 + 2\alpha\left(\omega - \omega_k^n\right)^2} \tag{32}$$

$$\omega_{k_{n+1}} = \frac{\int_0^\infty \omega \left| \hat{\mathrm{IMF}}_k(\omega) \right|^2 d\omega}{\int_0^\infty \left| \hat{\mathrm{IMF}}_k(\omega) \right|^2 d\omega} \tag{33}$$

$$\hat{\lambda}^{n+1}(\omega) = \hat{\lambda}^n(\omega) + \gamma \left(\hat{x}(\omega) - \sum_{k=1}^{K} \hat{\mathrm{IMF}}_k^{n+1}(\omega) \right) \tag{34}$$

Here, "^" refers to the Fourier transform operation. The "γ" refers to the noise tolerance. And the "n" refers to the number of iterations in the calculation process.

When the calculation result meets the given solution accuracy ε, the iterative operation is stopped and K IMFs are outputted [7].

$$\sum_{k=1}^{K} \left(\frac{\left\| \hat{\mathrm{IMF}}_k^{n+1} - \hat{\mathrm{IMF}}_k^{n} \right\|_2^2}{\left\| \hat{\mathrm{IMF}}_k^{n} \right\|_2^2} \right) < \varepsilon \tag{35}$$

Fifthly, the $\hat{\mathrm{IMF}}_{k_{n+1}}(\omega)$ is transformed from the frequency domain to the IMF in the time domain through the operation of the inverse Fourier transform.

Compared with the EMD, the VMD overcomes the problem of mode mixing and has a more solid mathematical theoretical foundation.

In the present chapter, the "K" is 7, the "α" is 2000, the "γ" is 0 with "ε" is 10^{-7}.

3.3 Examples of signal time-frequency analysis

This section uses the vibration signal at the top cover of a prototype reversible pump turbine during the start-up transient process of generating mode as the analysis object to demonstrate the applications of the time-frequency analysis methods introduced in Sections 3.1 and 3.2. The length of the signal is 2 s and the sampling frequency is 1000 Hz. **Figure 5** shows the time-domain diagram of the vibration signal at the top cover.

Figure 6 shows the result of the fast Fourier transform (FFT) of the vibration signal. It can be seen from the **Figure 6** that the most obvious frequency component in the vibration signal is 40 Hz with an amplitude of about 0.46 mm. In addition, there is also a relatively obvious peak near 80 Hz with an amplitude of about 0.17 mm.

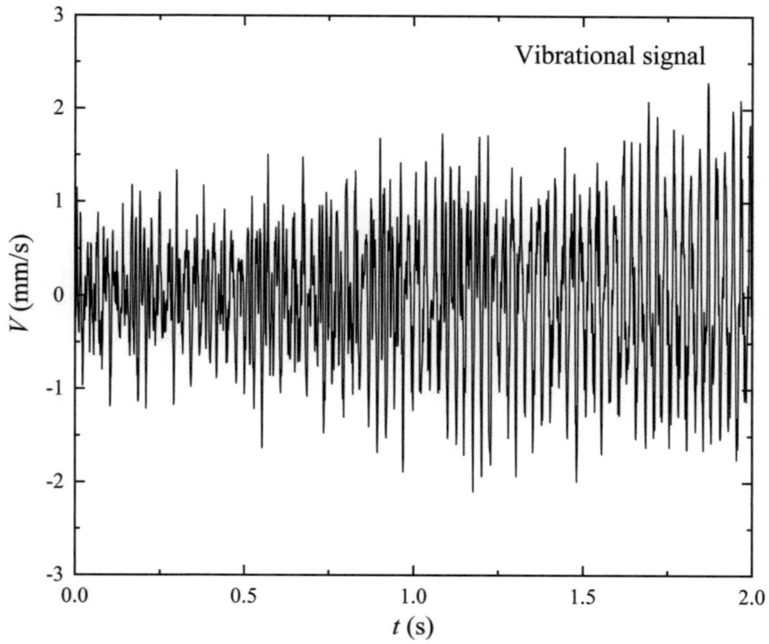

Figure 5.
The time-domain diagram of the vibration signal at the top cover.

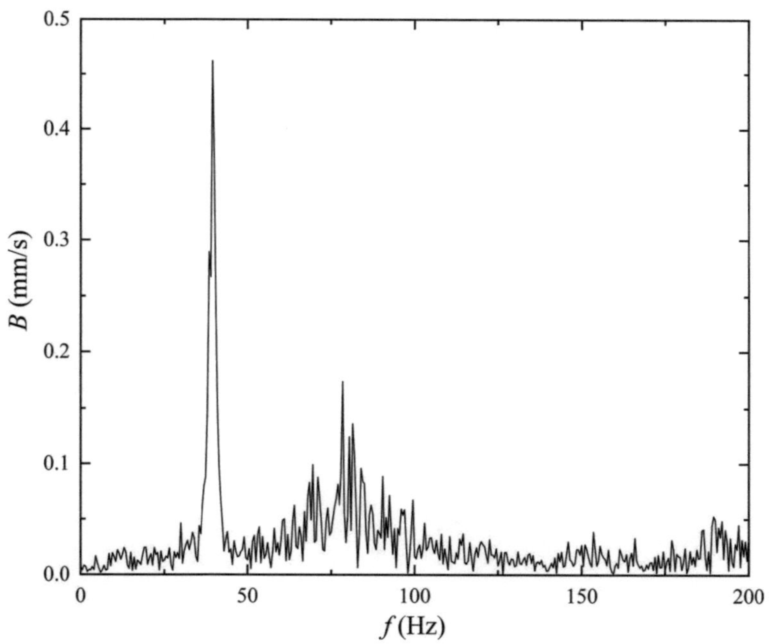

Figure 6.
The FFT spectrum of the vibration signal.

However, **Figure 6** cannot provide information about the amplitude variation of different frequency components with time, which is a shortcoming of the traditional Fourier transform.

Figures 7 and **8** are the STFT results of the vibration signal with the window length of 64 and 256 respectively. From these two figures, not only the amplitudes of different frequency components in the vibration signals but also the variations of the amplitude of each frequency varying with the time can be seen. However, the STFT results are significantly affected by the window size. By comparing **Figures** 7 and **8**, it can be concluded that there exist obvious amplitudes at and around 40 Hz in the **Figure** 7, while it almost has obvious amplitudes only at 40 Hz in the **Figure 8**, which is caused by the difference in the window size between them. In the actual application of the STFT, it is difficult to choose the appropriate window size, which is the main problem of this method.

Different with the fixed basis functions of traditional Fourier transform and STFT, HHT and VMD-Hilbert transform can obtain basis functions through adaptive mode decomposition. In addition, HHT and VMD-Hilbert transform do not need to select the window functions and they can achieve high resolutions both in time and frequency domains. **Figures 9** and **10** are the results of the HHT and VMD-Hilbert

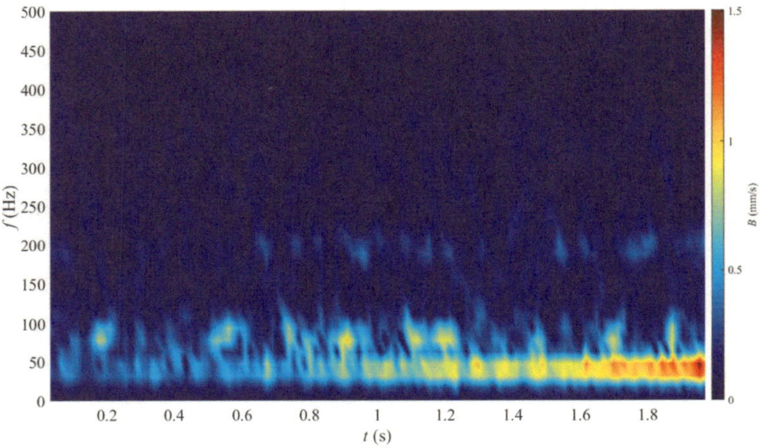

Figure 7.
The STFT result of the vibration signal with the window length of 64 and the window function of Blackman.

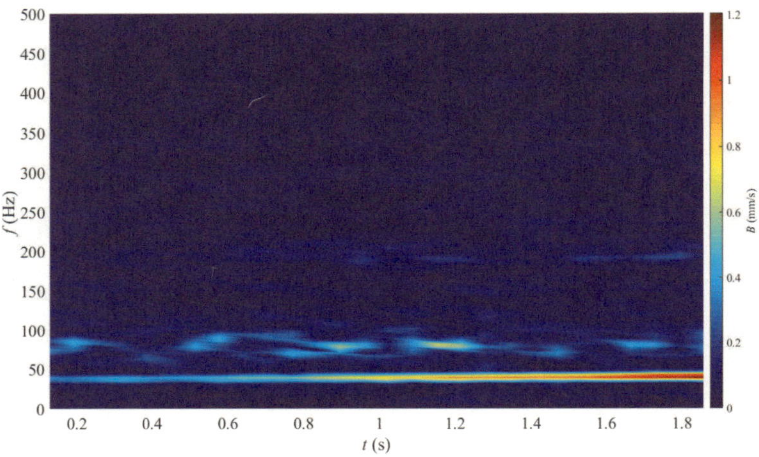

Figure 8.
The STFT result of the vibration signal with the window length of 256 and the window function of Blackman.

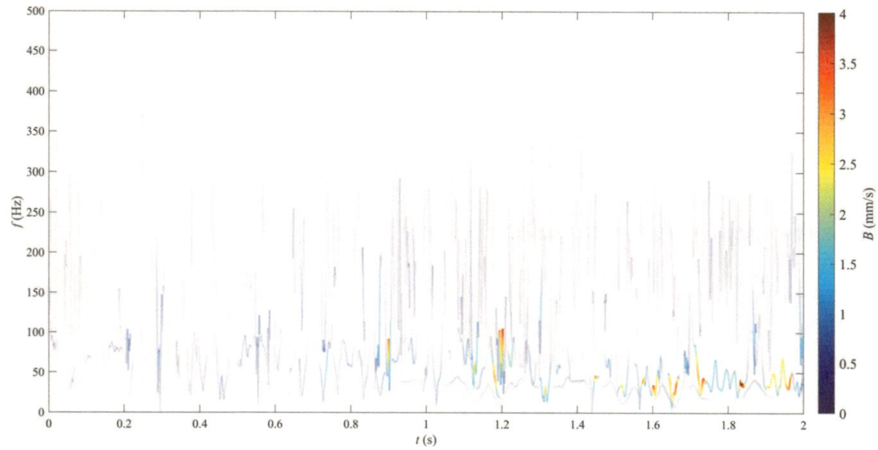

Figure 9.
The analysis result of HHT of the vibration signal.

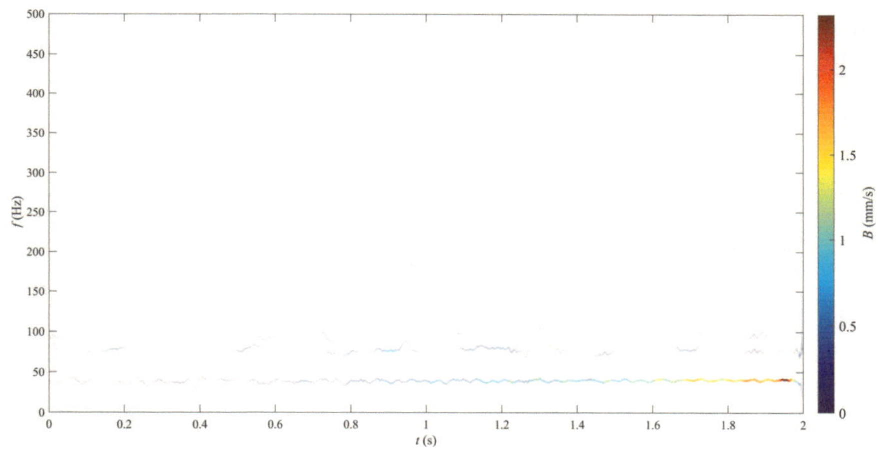

Figure 10.
The analysis result of VMD-Hilbert transform of the vibration signal.

transform of the signal, respectively. It can be found that the analysis result in the **Figure 9** is unclear because the dominant frequencies of the signal cannot be clearly observed, while the analysis result in the **Figure 10** is very clear. This can be attributed to that the VMD overcomes the mode mixing problem in the EMD adopted in the HHT.

4. Signal mode decomposition and signal de-noising

Mode refers to the natural vibration characteristics of the mechanical structure, which is composed of natural frequency, damping ratio and mode shape. In practical engineering, the vibration of rotating machinery is often composed of different complex modes. The process of decomposing the vibration signal into sub-signals with different frequency bands on the basis of different modes is called the mode decomposition of the signal. Through the mode decomposition, the modes containing noise can be found and will be further deleted through the signal de-noising. In our previous research [8], similar methods with those shown in this section have been adopted to analyze the vibration signal from the rotating machinery of nuclear power. However, the specific frequency components between the signals between the rotating machinery of nuclear power and the reversible pump turbine are a little different.

4.1 Mode decomposition of simulated swing signal

Typical signal analysis methods, such as EMD and VMD, have been fully described in Section 3.2. In this section, a simulated swing signal at the turbine guide bearing of a prototype pump turbine will be used to demonstrate the signal de-noising effects based on these two methods. The simulated rotational speed of the unit is 500 rpm. The characteristic frequencies contained in the simulated signal are 1 time, 2 times and 3 times, the rotational frequency of the impeller with the corresponding amplitudes of 10, 2 and 1 μm, respectively. The signal is dimensionless by dividing the corresponding amplitude of the rotational frequency of the impeller (10 μm). The length of the signal is 2 s and the sampling frequency is 1000 Hz. The white noise with the signal-noise ratio (*SNR*) of 20 dB is added to the simulated swing signal. **Figure 11** shows the time-domain diagrams of the simulated swing signals before and after

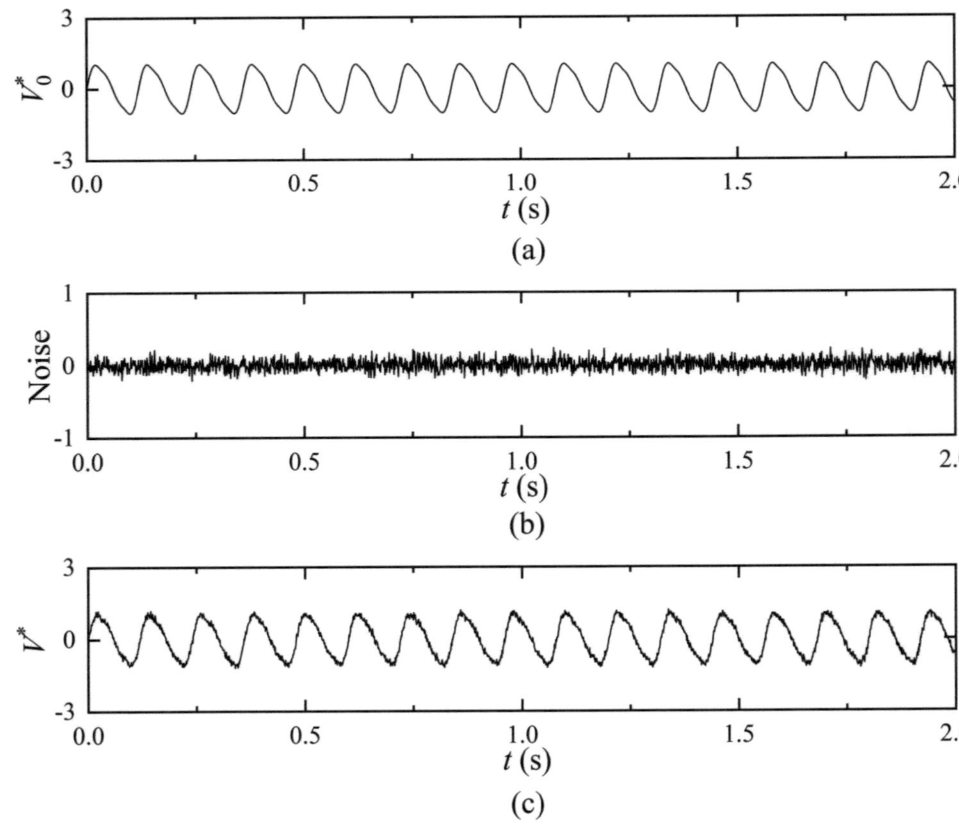

Figure 11.
The time-domain diagrams of simulated swing signals. (a) The simulated swing signal without noise. (b) The added noise with the SNR *of 20 dB. (c) The simulated swing signal with noise added.*

adding the white noise. **Figures 12** and **13** are the EMD and VMD analysis results of the simulated swing signal, respectively.

4.2 Results of signal de-noising

Through mode decomposition, several IMFs are obtained to further constitute the given signal with low noise. The IMFs contain different frequency bands. By removing

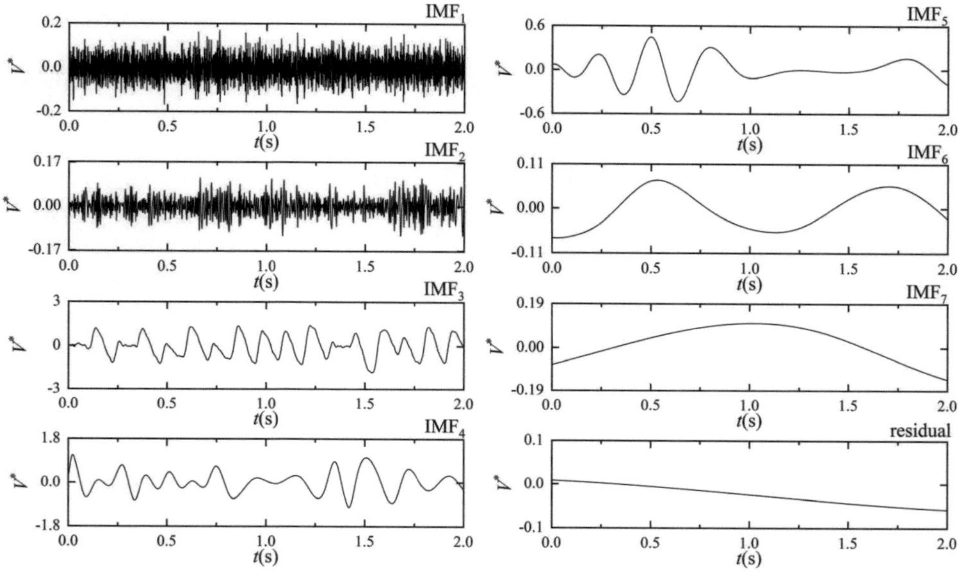

Figure 12.
The EMD analysis results of the simulated swing signal.

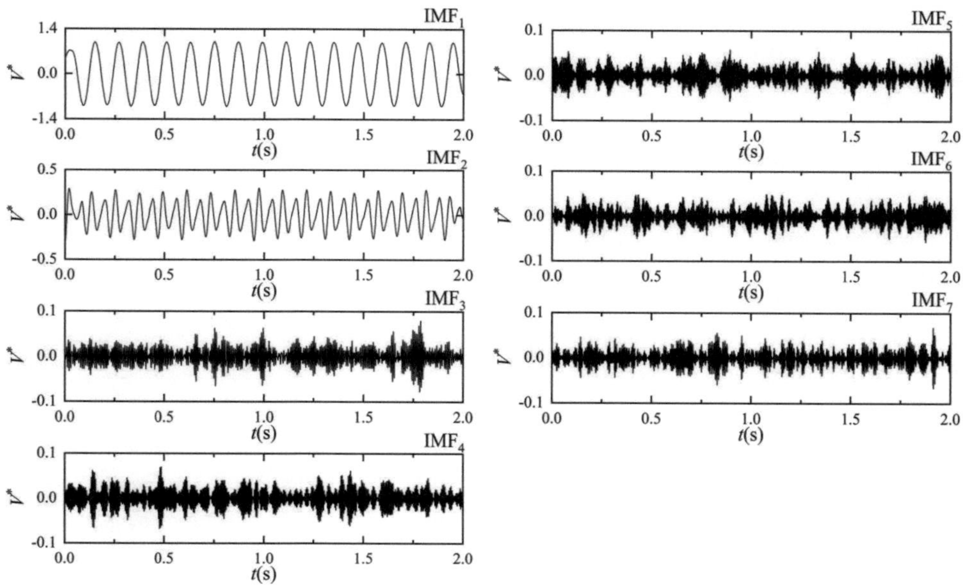

Figure 13.
The VMD analysis results of the simulated swing signal.

the IMFs dominated by the random noises and retaining the IMFs dominated by the useful signals, the signal de-noising will be proceeded. Generally speaking, the cross-correlation coefficient between the noise-dominated IMF and the original signal is small, while the cross-correlation coefficient between the IMF component dominated by the useful signal and the original signal is relatively large. Based on this principle, the noise-dominated IMFs can be identified according to the cross-correlation coefficients.

Table 3 shows the cross-correlation coefficients of the IMFs obtained by EMD and VMD and the simulated swing signal. Here, the threshold of the cross-correlation coefficient is set to 0.12 [8]. And the IMFs with the cross-correlation coefficients larger than this threshold are retained.

Based on the analysis results in **Table 3**, the IMF_3 and IMF_4 in the EMD analysis results and the IMF_1 and IMF_2 in the VMD analysis results are selected to reconstruct the de-noised swing signals, respectively. **Figure 14** shows the de-noising results of the simulated swing signal based on EMD and VMD, respectively.

Figure 14 shows that both the de-noised signals based on EMD and VMD are relatively smooth oscillation curves. Compared with the simulated swing signal without noise, the de-noised signal based on EMD has some abnormal spikes or irregularities as indicated by the red circles in the **Figure 14b**, while the time-domain waveform of the de-noised signal based on VMD is much closer to the simulated swing signal without noise, which indicates that the signal de-noising based on VMD has a better effect than that based on EMD.

4.3 Evaluation indexes of signal de-noising

The evaluation indexes of signal de-noising can help us quantitatively analyze the de-noising effects. Some evaluation indexes will be introduced below. Here $x(t)$ refers to original simulated swing signal without noise; "i" refers to the i-th data point of signal $x(t)$ which contains M data points in total; $\tilde{x}_i$refers to de-noised swing signal.

Signal-noise ratio (SNR)

$$SNR = 10\lg\left(\frac{\sum_{i=1}^{M} x_i^2}{\sum_{i=1}^{M} (x_i - \tilde{x}_i)^2}\right) \tag{36}$$

The larger the SNR, the better the de-noising effect.

Mode component	EMD	VMD
IMF_1	0.0757	0.9732
IMF_2	0.0537	0.2378
IMF_3	0.7777	0.0453
IMF_4	0.3762	0.0430
IMF_5	0.0067	0.0407
IMF_6	0.0242	0.0412
IMF_7	0.0349	0.0400
residual	0.0103	—

Table 3.
Cross-correlation coefficients of IMFs based on EMD and VMD and the simulated swing signal.

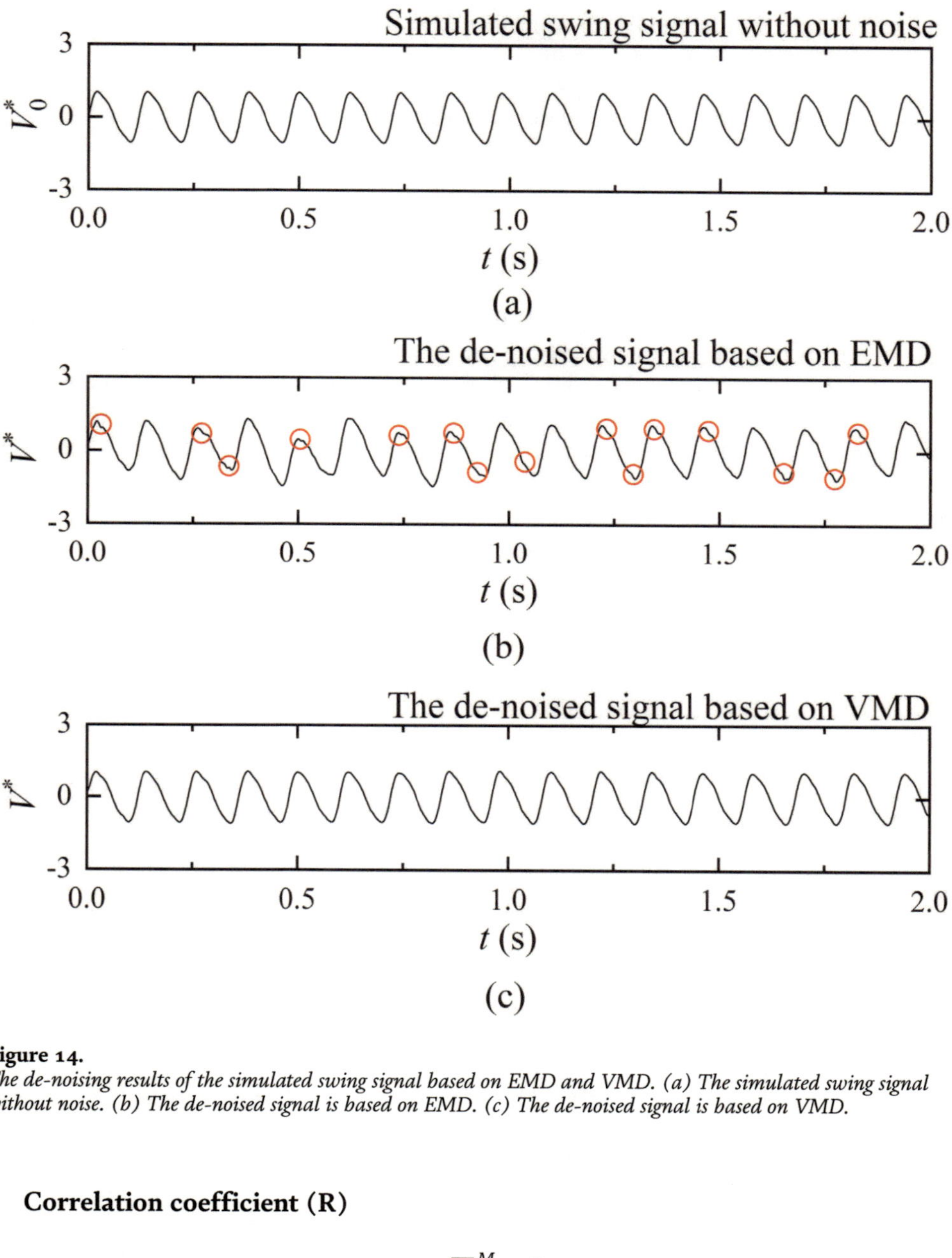

Figure 14.
The de-noising results of the simulated swing signal based on EMD and VMD. (a) The simulated swing signal without noise. (b) The de-noised signal is based on EMD. (c) The de-noised signal is based on VMD.

Correlation coefficient (R)

$$R = \frac{\sum_{i=1}^{M} x_i \tilde{x}_i}{\sqrt{\sum_{i=1}^{M} x_i^2 \sum_{i=1}^{M} \tilde{x}_i^2}} \tag{37}$$

The larger the R, the better the de-noising effect.
Root-mean-square error (RMSE)

$$RMSE = \sqrt{\frac{1}{M} \sum_{i=1}^{M} (x_i - \tilde{x}_i)^2} \tag{38}$$

The smaller the RMSE, the better the de-noising effect.

The evaluation indexes of signal de-noising effects based on EMD and VMD are calculated respectively and the results are shown in **Table 4**.

It can be seen from **Table 4** that compared with the results of signal de-noising based on EMD, the results of signal de-noising based on VMD have a larger *SNR*, a larger *R*, and a smaller *RMSE*, which further shows that the signal de-noising based on VMD has a better effect than that based on EMD.

4.4 Decomposition and de-noising of measured swing signal

In Sections 4.1–4.3, the simulated swing signal is adopted to demonstrate the signal decomposition and the signal de-noising processes. In this section, the analysis results of mode decomposition and de-noising of measured swing signal of a prototype pump turbine are presented. The swing signal at the turbine guide bearing of a pump turbine was measured at P^*=15%. The rated rotational speed of the unit is 500 rpm. The length of the signal is 2 s and the sampling frequency is 1000 Hz. **Figure 15** shows the time-domain diagram of the measured swing signal.

Evaluation index	EMD	VMD
SNR	12.1328	29.5778
R	0.969	0.9995
RMSE	0.1856	0.024

Table 4.
The evaluation indexes of de-noising for the simulated swing signal based on EMD and VMD.

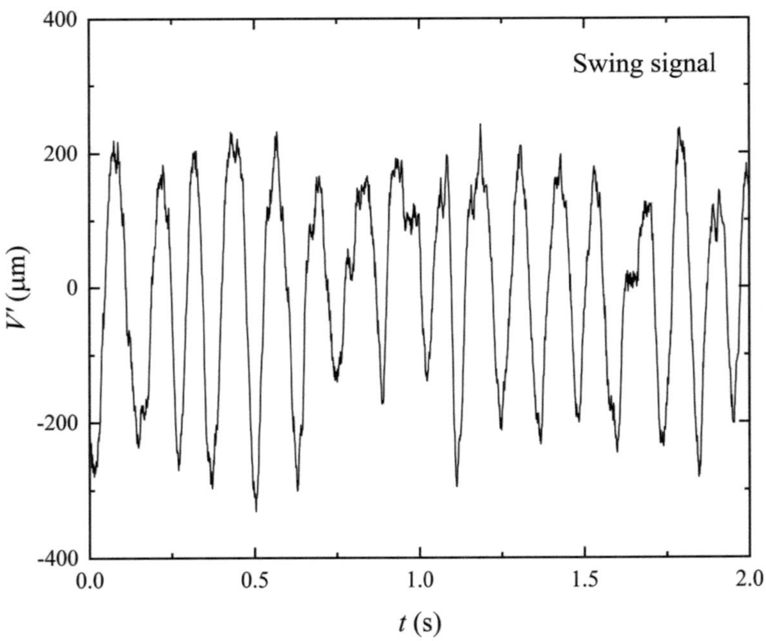

Figure 15.
The time-domain diagram of the measured swing signal.

The EMD and VMD analysis results of the measured swing signal are shown in **Figures 16** and **17**.

The cross-correlation coefficients of the IMFs obtained by EMD and VMD and the measured swing signal are shown in **Table 5**.

The threshold of the cross-correlation coefficient is also set to 0.12 [8]. Based on the analysis results in **Table 5**, the IMF_3, IMF_4, and IMF_7 in the EMD analysis results

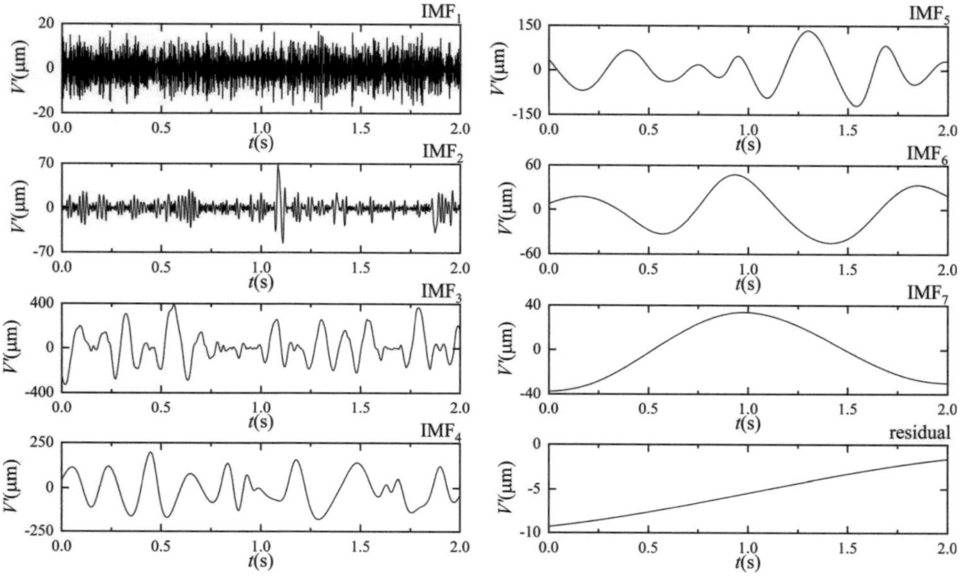

Figure 16.
The EMD analysis results of the measured swing signal.

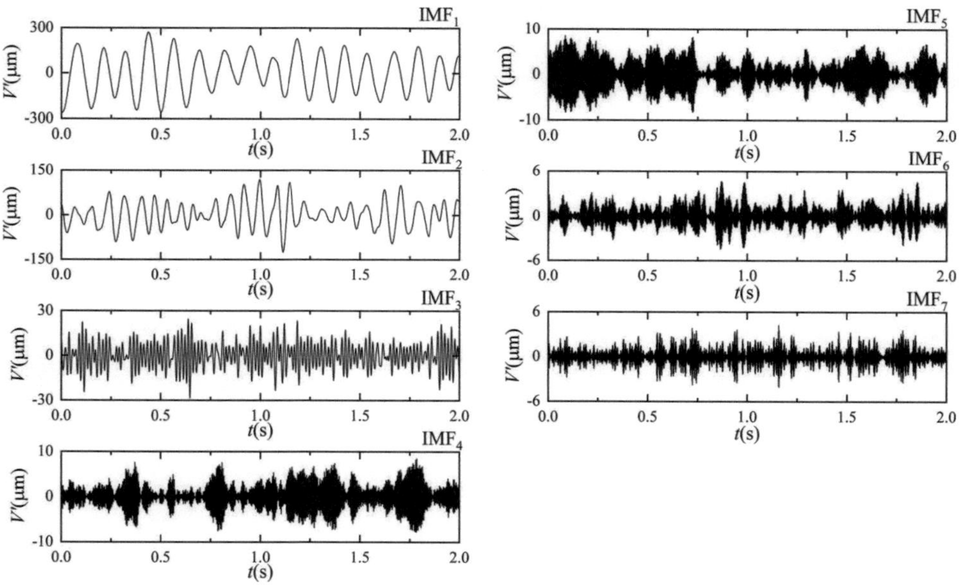

Figure 17.
The VMD analysis results of the measured swing signal.

Mode component	EMD	VMD
IMF_1	0.0635	0.9471
IMF_2	0.0797	0.4739
IMF_3	0.7677	0.1096
IMF_4	0.2904	0.0323
IMF_5	0.0500	0.0321
IMF_6	0.0715	0.0200
IMF_7	0.1455	0.0154
residual	0.0444	—

Table 5.
The cross-correlation coefficients of the IMFs based on EMD and VMD and the measured swing signal.

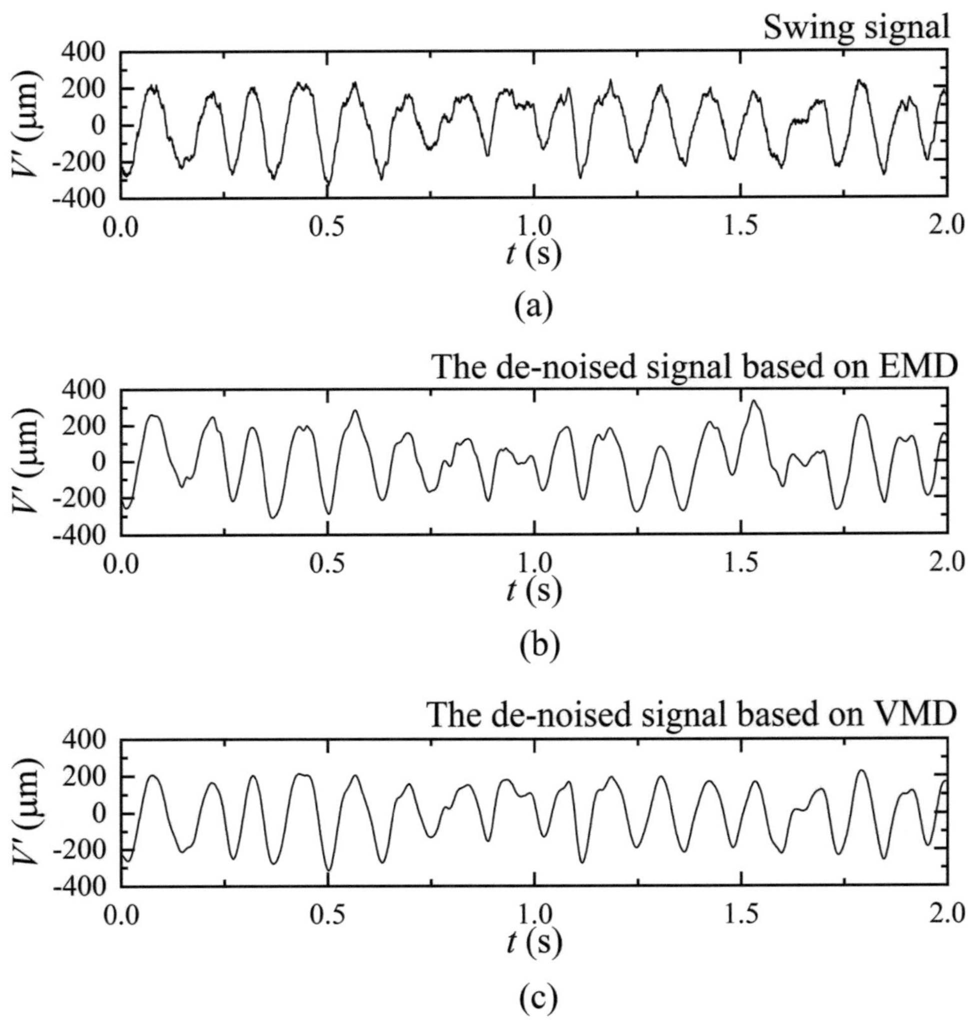

Figure 18.
The de-noising results of the measured swing signal is based on EMD and VMD. (a) The original measured swing signal. (b) The de-noised signal is based on the EMD. (c) The de-noised signal based on VMD.

and the IMF_1 and IMF_2 in the VMD analysis result are selected to reconstruct the de-noised swing signals, respectively. **Figure 18** shows the de-noising results of the measured swing signal based on EMD and VMD, respectively. As shown in **Figure 18**, the VMD shows a better de-noising effect by comparing the smoothness of the curves.

5. Conclusions

In summary, signal analysis and processing can help us better understand the operational states of the reversible pump turbines. Through the time-domain analysis, the time-domain characteristic parameters such as peak value, peak-to-peak value and average value can be obtained. Through the time-frequency analysis, the time-frequency variation characteristics of signals can be grasped. Through the mode decomposition of the signal, various components within the signal can be distinguished and the signal de-noising can be further performed. Therefore, signal analysis plays a very important role for the condition monitoring and fault diagnosis of the reversible pump turbines.

Acknowledgements

This work was financially supported by the National Natural Science Foundation of China (Project Nos.: U1965106, 51976056 and 52076215).

Author details

Hao Li[1], Xianghao Zheng[1], Yuning Zhang[1*], Jinwei Li[2] and Yuning Zhang[3]

1 North China Electric Power University, Beijing, China

2 China Institute of Water Resources and Hydropower Research, Beijing, China

3 China University of Petroleum, Beijing, China

*Address all correspondence to: yuning.zhang@foxmail.com

References

[1] Zhang Y, Zheng X, Li J, et al. Experimental study on the vibrational performance and its physical origins of a prototype reversible pump turbine in the pumped hydro energy storage power station. Renewable Energy. 2019;**130**: 667-676. DOI: 10.1016/j.renene. 2018.06.057

[2] Zhang Y, Chen T, Li J, et al. Experimental study of load variations on pressure fluctuations in a prototype reversible pump turbine in generating mode. ASME Journal of Fluids Engineering. 2017;**139**(7):074501. DOI: 10.1115/1.4036161

[3] Zheng X, Zhang Y, Li J, et al. Influences of rotational speed variations on the flow-induced vibrational performance of a prototype reversible pump turbine in spin-no-load mode. ASME Journal of Fluids Engineering. 2020;**142**(1):011106. DOI: 10.1115/1.4045159

[4] Bandt C, Pompe B. Permutation entropy: a natural complexity measure for time series. Physical review letters. 2002;**88**(17):174102. DOI: 10.1103/ PhysRevLett.88.174102

[5] Gabor D. Theory of communication. Part 1: The analysis of information. Journal of the Institution of Electrical Engineers-Part III: Radio and Communication Engineering. 1946; **93**(26):429-441

[6] Huang NE, Shen Z, Long SR, et al. The empirical mode decomposition and the Hilbert spectrum for nonlinear and non-stationary time series analysis. Proceedings Mathematical Physical & Engineering Sciences. 1971;**1998**(454): 903-995. DOI: 10.1098/rspa.1998.0193

[7] Dragomiretskiy K, Zosso D. Variational mode decomposition. IEEE Transactions on Signal Processing. 2013; **62**(3):531-544. DOI: 10.1109/TSP.2013. 2288675

[8] Zheng X, Zhang Y, Li J, et al. Research on the signal de-noising of the rotating machinery based on the variational mode decomposition and the cross-correlation analysis. Nuclear Science and Engineering. 2022. (in press)

Chapter 5

SOFC-Gas Turbine Hybrid Power Plant: Exergetic Study

Salha Faleh and Tahar Khir

Abstract

The combined solid oxide fuel cells and gas turbine hybrid system is known to be a promise alternative for power generation with high efficiency. This paper presents the third part of the parametric study of a Solid Oxide Fuel Cell/Gas Turbine (SOFC/GT) hybrid system generating 120 MW Net power. The studied parameters are Pressure *P*, pre-reformed fraction X_r, extraction fraction fs, H_2 flow, and air flow. Their effects on the performances of SOFC-GT hybrid system are investigated. The Engineering Equations Solver (EES) simulation is established to analyze the SOFC-GT exergetic and energetic system performances. The results show that increasing the air and fuel flows enhance system exergy. In contrary, the pressure at the SOFC and the extraction fraction negatively affects the exergy performance of the hybrid plant. It is also found that the combustion chamber, pre-reformer, and SOFC represent the greatest exergy destroyers.

Keywords: gas turbine, SOFC, exergetic efficiency, parametric study

1. Introduction

Power generation with high efficiency and low emissions becomes a serious research topic. Fuel cells are considered as an appropriate technology. Among several fuel cells, Solid Oxide Fuel Cell (SOFC) is characterized by special advantages that make its coupling with different power cycles especially with Gas Turbine (GT) an attractive proposal to achieve high electrical and thermal efficiencies. Many papers studied SOFC/GT hybrid systems. This integration shows a high electrical efficiency [1]. Parametric, economic, energetic, and exergetic studies are established to optimize SOFC/GT systems [2, 3]. Also, the advancement in material technology and its application on SOFC [4], the use of different fuels, modeling techniques, various designs, techniques, and configurations have been proposed and studied [5–13]. Additional electrochemical systems, bottoming cycles, and chemical processes can be integrated in SOFC-Turbine hybrid system to enhance its efficiency such as the organic Rankine bottom cycle with cryogenic nitrogen assisting in CO_2 recovery used by Yang et al. [14]. A multi-generation system power investigated by Haghghi et al. gave 47.14% as an overall system exergy efficiency [15]. The solar array, the proton exchange membrane electrolyzer, and digester are added by Inac et al. to develop a highly renewable hybrid concept [16]. Gholamian et al. [17] found that ORC integrated with

IntechOpen

thermoelectric generator is having high exergy efficiency in comparison of the basic ORC by 21.9%. Malico et al. [18] studied the integrated SOFC-absorption chiller system for cooling, heating, and power by utilizing waste heat from SOFC and found 68% thermal efficiency. Mehrpooya et al. [19] investigated SOFC-Ammonia water single effect absorption system with ORC combination and they found an efficiency of 62.4%.

Different architectures of hybrid systems are described in previous research. Coupling SOFCs thermally or chemically with bottoming power cycles and renewable energy conversion and storage devices achieves an important efficiency reaching 80% [20–23]. Valerie et al. [24] assessed a thermo-economic study of an indirect integration of a standard gas turbine cycle with an internal reforming SOFC system and bottoming Organic Rankine Cycle (ORC). They concluded that with toluene as working fluid an energy efficiency of about 4% and an exergy efficiency of about 62% are obtained. A supercritical CO_2 bottoming cycle was added by Meng et al. to operate the SOFC at higher power density with 70% system efficiency [25]. Many combinations show an important efficiency. Liu et al. found that 62.29% exergy efficiency was obtained using SOFC-GT-ORG-CO_2 CAPTURE [26]. An overall exergy efficiency of 47.14% was obtained by a multi-generation system for power, cooling, and hydrogen and water production [27]. An electrical efficiency of about 75.8% is obtained by Yi et al. [28]. An improvement of the exergy and the energy performances by 26.6 and 27.8% of SOFC-GT hybrid system compared to traditional gas turbine performances, respectively, is demonstrated by Haseli et al. [29]. A thermodynamic and exergo-economic analysis of a SOFC-GT cogeneration system were conducted by Mahmoudi and Khani [2]. They found that the economic and thermodynamic performances increase with the TIT and steam to carbon ratio. Harvey and Richter [30] used ASPEN plus simulator to study pressurized SOFC system, initially developed in Argonne National Laboratory. Later on in the 1997, Siemens Westinghouse delivered the first prototype for producing 220 kW via pressurized SOFC-GT. Santhanam et al. [6] studied an SOFC-GT system integrated to a bio-mass gasifier while integrating a heat pipe within the SOFC stack. They found reduction in exergy destruction and improvement in efficiency of the proposed system. The electric efficiency achieved for the proposed plant exceeds 72%. Ji et al. [10] compared performance assessment of integration of hybrid SOFC-GT cycle with different bottoming Rankine cycle alternatives. They concluded that triple combined cycle yields 3% higher efficiency as compared to dual combined cycle.

2. System description

The studied power plant is represented by **Figure 1**. It is mainly composed by a gas turbine cycle GT, a Solid Oxide Fuel Cell SOFC system, and an ammonia water absorption refrigerating system; also heat exchangers HE is used to recover and valorize the heat fluxes from the Gas Turbine and SOFC exhausts. Before feeding the SOFC, the fuel is pre-reformed for a specific portion. The required steam for pre-reforming is produced by the steam generator working with the exhaust heat 43 g from HE1. The outlet of the performing 3C will be completely reformed in the SOFC. After the compression through CA2 the ambient air 3a will be heated in HE1 to obtain 4C. Inside the SOFC, the chemical energy will be converted to electricity due to electrochemical reaction that occurs in the SOFC. Fuel cell exhaust 43a is compressed to 43c to be used to heat the air compressed by CA1 in the heat exchanger HE2. The fuel mixture (the gas 43d and the fresh fuel 2c) will feed the combustion chamber

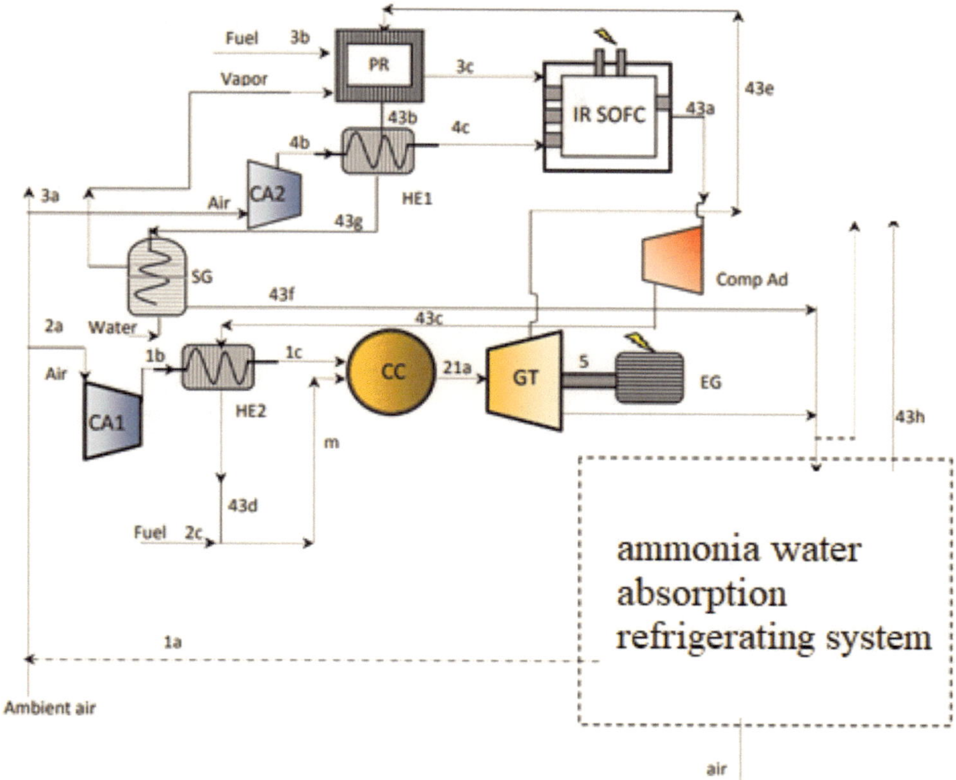

Figure 1.
SOFC-GT hybrid system power plant diagram.

with the compressed heated air 1c. The exhaust flow from the combustion chamber 21a is expanded by the gas turbine GT to generate mechanical power which is converted in electrical generator to electricity. In addition, the flux 43f is used to activate the ammonia water absorption refrigerating system. The exhaust flux from the GT is transferred to the refrigerating system or to the stack.

3. Mathematical models

The purpose of exergy analysis is to determine the performance and identify the sources of irreversibility of the hybrid system.

3.1 SOFC

Characteristics of the fuel cell are selected according to ref. [31]. The SOFC is the seat of reforming, shifting, and electrochemical reactions. The used fuel composition is reported in **Table 1**. Developed SOFC model calculates voltages, power, and outlet flow parameters.

The power delivered by the SOFC is given by:

$$P_{SOFC} = N_p * E_{feel} * I$$

Compound	Molar composition (%)
N_2	3.04
CO_2	0.64
CH_4	78.95
C_2H_6	13.5
C_3H_8	3.87

Table 1.
Natural gas molar composition.

The exergy balance and the exergetic efficiency of the SOFC can be expressed as:

$$\dot{E}_{3c} + \dot{E}_{4c} = \dot{E}_{43a} + \dot{w}_{SOFC} + \dot{ED}_{SOFC}$$

$$\varepsilon_{SOFC} = \frac{\dot{w}_{SOFC}}{\dot{E}_{4c} + \dot{E}_{3c} - \dot{E}_{43a}}$$

3.2 Gas turbine

The gas turbine outlet temperature is given by

$$T_{5s} = T_{21a}/R^{\left(1-\frac{1}{k}\right)};$$

Where R is the gas turbine expansion ratio

$$R = \frac{P_{21a}}{P_{5s}}$$

The power produced by the GT is calculated as follows:

$$\dot{W}_{TG} = \eta_t * \left[\left(\dot{H}_{21a} - \dot{H}_{43e}\right) + (1 - St) * \left(\dot{H}_{43e} - \dot{H}_{5s}\right)\right]$$

The exergy balance and the exergetic efficiency of the GT can be expressed as:

$$\dot{E}_{21a} = \dot{E}_5 + \dot{w}_{TG} + \dot{ED}_{TG}$$

$$\varepsilon_{TG} = \frac{\dot{w}_{TG}}{\dot{E}_{21a} - \dot{E}_5}$$

3.3 Combustion chamber

The exergy balance and the exergetic efficiency of the combustion chamber can be expressed as:

$$\dot{E}_{2c} + \dot{E}_m = \dot{E}_{21a} + \dot{ED}_{CC}$$

$$\varepsilon_{cc} = \frac{\dot{E}_{21a}}{\dot{E}_{2c} + \dot{E}_m}$$

3.4 Heat exchangers

Multiple gas-to-gas heat exchangers are proposed to recover heat. It is assumed that there is no heat transfer between these elements and the surrounding environment. The effectiveness- NTU method is used to determine the actual temperature changes for both cold and hot fluids, based on the heat exchanger type, effective heat transfer coefficient, and surface area. For a cross flow and unmixed fluid type heat exchanger, the effectiveness is expressed as [1]:

$$\epsilon = 1 - \exp\left\{\frac{\mathrm{NTU}^{0,22}}{\mathrm{Cr}}\left[\exp\left(-\mathrm{Cr.NTU}^{0,78}\right) - 1\right]\right\}$$

$$\mathrm{C_r} = \frac{C_{\min}}{C_{\max}}$$

$$C_{\min} = \mathrm{Minimum}\left(\dot{m}_c C p_c, \dot{m}_f C p_f\right)$$

$$\mathrm{NTU} = \frac{\mathrm{UA}}{C_{\min}}$$

The exergy balance and the exergetic efficiency of the heat exchangers are determined using the following equations:

$$\dot{\mathrm{E}}_{1b} + \dot{\mathrm{E}}_{43c} = \dot{\mathrm{E}}_{1c} + \dot{\mathrm{E}}_{43d} + \dot{\mathrm{ED}}_{\mathrm{ECTG}}$$

$$\varepsilon_{\mathrm{ECGT}} = \frac{\dot{\mathrm{E}}_{1c} - \dot{\mathrm{E}}_{1b}}{\dot{\mathrm{E}}_{43c} - \dot{\mathrm{E}}_{43d}}$$

$$\dot{\mathrm{E}}_{3b} + \dot{\mathrm{E}}_{43b} = \dot{\mathrm{E}}_{3c} + \dot{\mathrm{E}}_{43g} + \dot{\mathrm{ED}}_{\mathrm{ECSOFC}}$$

$$\varepsilon_{\mathrm{ECSOFC}} = \frac{\dot{\mathrm{E}}_{3c} - \dot{\mathrm{E}}_{3b}}{\dot{\mathrm{E}}_{43b} - \dot{\mathrm{E}}_{43g}}$$

3.5 Compressors

The outlet temperatures of used compressors are calculated using the following equations:

$T_{1b} = T_{1a} * R_1^{\left(1-\frac{1}{k}\right)}$ Where $R_1 = \frac{P_{1b}}{P_{1a}}$ is the compression ratio of $\mathrm{comp_{GT}}$.

$T_{3b} = T_{3a} * R_3^{\left(1-\frac{1}{k}\right)}$ Where $R_3 = \frac{P_{3b}}{P_{3a}}$ is the compression ratio of $\mathrm{comp_{SOFC}}$.

The power consumed by different compressors can be calculated using the following equations:

$$\dot{\mathrm{W}}_{\mathrm{comp\ TG}} = \frac{\left(\dot{\mathrm{H}}_{1a} - \dot{\mathrm{H}}_{1b}\right)}{\eta_c}$$

$$\dot{\mathrm{W}}_{\mathrm{comp\ TG}} = \frac{\left(\dot{\mathrm{H}}_{3a} - \dot{\mathrm{H}}_{3b}\right)}{\eta_c}$$

$$\dot{\mathrm{W}}_{\mathrm{comp\ supp}} = \frac{\left(\dot{\mathrm{H}}_{43c} - \dot{\mathrm{H}}_{43a}\right)}{\eta_c}$$

The compressors' exergy balance and the exergetic efficiency are determined using following equations:

$$\dot{E}_{1a} + \dot{w}_{compTG} = \dot{E}_{1b} + \dot{ED}_{compTG}$$

$$\varepsilon_{compTG} = \frac{\dot{E}_{1a} - \dot{E}_{1b}}{\dot{w}_{compTG}}$$

$$\dot{E}_{3a} + \dot{w}_{compSOFC} = \dot{E}_{3b} + \dot{ED}_{compSOFC}$$

$$\varepsilon_{compTG} = \frac{\dot{E}_{3a} - \dot{E}_{3b}}{\dot{w}_{compSOFC}}$$

$$\dot{E}_{43a} + \dot{w}_{compsup} = \dot{E}_{43c} + \dot{ED}_{compsupp}$$

$$\varepsilon_{compsupp} = \frac{\dot{E}_{43a} - \dot{E}_{43c}}{\dot{w}_{compsupp}}$$

3.6 Overall cycle efficiency

The net system power of hybrid system is given by

$$P_{NET} = P_{tot} - \dot{W}_{compTG} - \dot{W}_{compSOFC} - \dot{W}_{compsupp}$$

The energetic efficiency of gas turbine cycle is

$$\eta_{en,TG} = \frac{\dot{W}_{TG} - \dot{W}_{comp\ TG}}{\dot{m}_{fGT} * PCI}$$

The energetic efficiency of SOFC is

$$\eta_{en,SOFC} = \frac{P_{SOFC} - \dot{W}_{compSOFC}}{\dot{m}_{fSOFC} * PCI}$$

The energetic efficiency of the hybrid system is

$$\eta_{en,sys} = \frac{P_{NET}}{\left(\dot{m}_{fSOFC} + \dot{m}_{fGT}\right) * PCI}$$

4. Results and interpretations

1. Effect of pressure

The air compressor consumes a lot of energy. The more the compression ratio increases, the more the energy requirement of this component increases. So, as the cell pressure increases, the compressor exergy decreases. Moreover, the pressure has a negative effect on the exergy efficiency of the fuel cell and the gas turbine as shown in **Figures 2** and **3**. Operating at a pressure equal to 200 kPa, the cell and the turbine exhibit exergy efficiencies of approximately 57.3 and 80.7%, respectively. Whereas for a pressure P = 1000 kPa, these efficiencies

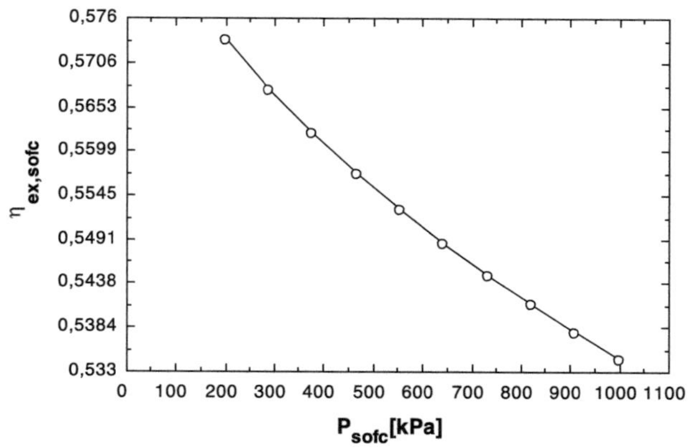

Figure 2.
Effect of pressure on SOFC exergetic efficiency.

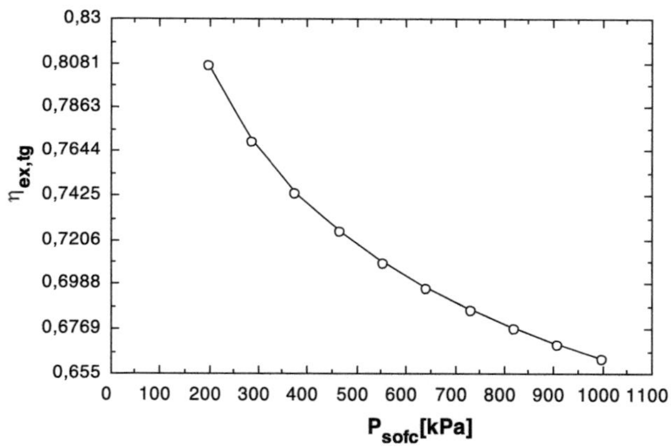

Figure 3.
Effect of pressure on gas turbine exergetic efficiency.

become respectively 53.4 and 66.2%. In addition, the pressure variation has a small influence on the exergy efficiency of the heat exchanger of the gas turbine cycle and a large effect on the exergy efficiency of the heat exchanger of the SOFC cycle as shown in **Figure 4**.

2. Effect of the extraction fraction

The increase in the extraction fraction implies a decrease in the useful power. The turbine loses part of the load to be expanded by the extraction leading to a decrease in the power generated as illustrated in **Figure 5**. An extraction of 4.5% leads to an efficiency of 67.4%, while a 65.8% yield is obtained if the extraction is doubled. The effect of the extracted fraction on the exergy of the stack heat exchanger was also studied as shown in **Figure 6**. By increasing the extracted fraction, the exergy of the exchanger decreases. This decrease is due to the increase in energy flows due to the increase in fluid mass flow. The more the mass flow increases, the greater the irreversibility. For extracted fractions of 0.45 and 0.8, the exergy yields obtained are 82% and 79%, respectively. Furthermore, a practically negligible negative effect of the extracted fraction on

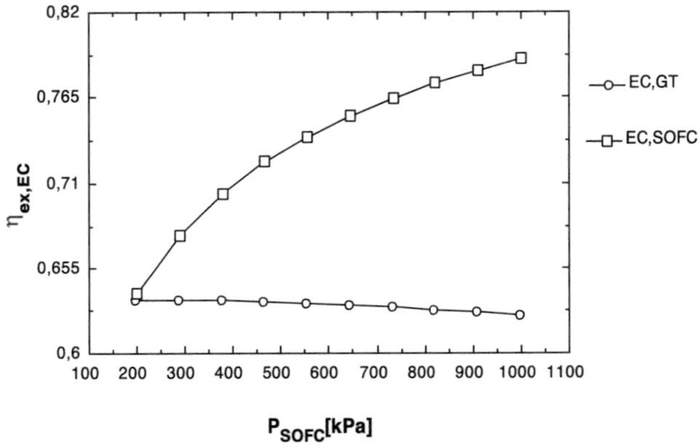

Figure 4.
Effect of pressure on heat exchangers' exergetic efficiencies.

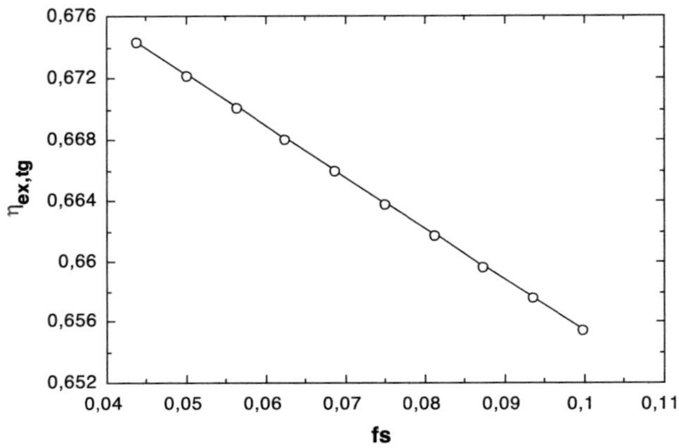

Figure 5.
Effect of extraction fraction on the gas turbine exergetic efficiency.

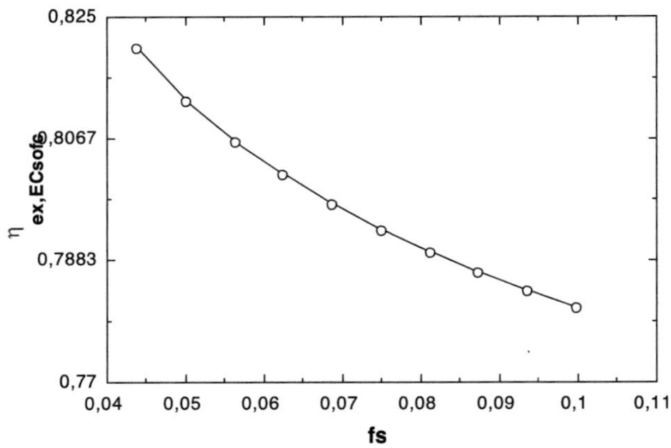

Figure 6.
Effect of the extraction fraction on the exergetic efficiency of SOFC-Heat exchanger.

the fuel cell is observed, as shown in **Figure 7**. Indeed, by increasing from a 5% extraction fraction to 10%, the destruction of exergy increases by only 2%. However, this parameter has a positive effect on the pre-reformer. **Figure 8**

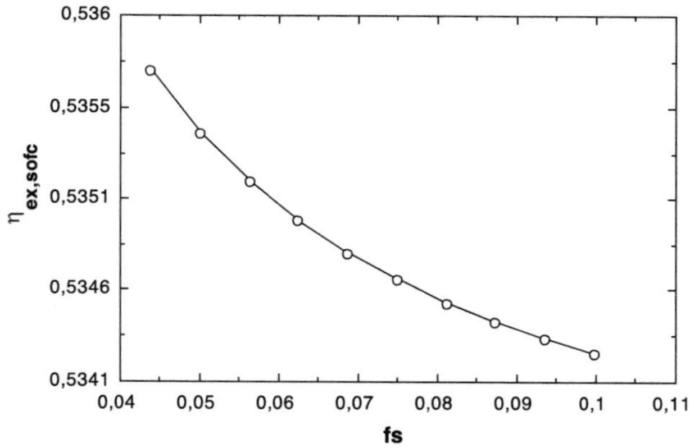

Figure 7.
Effect of the extraction fraction on the SOFC exergetic efficiency.

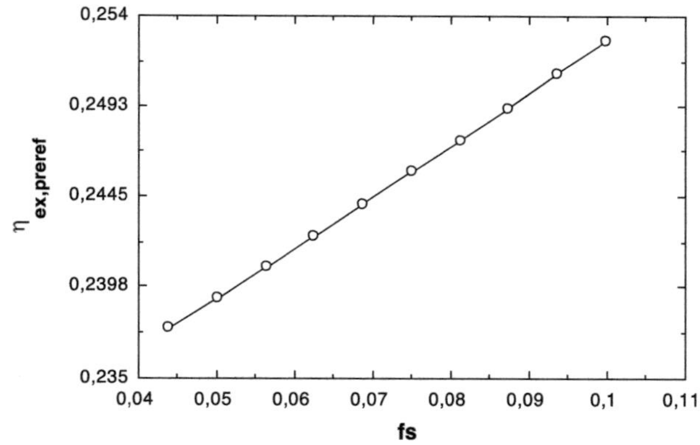

Figure 8.
Effect of the extraction fraction on the pre-reformer exergetic efficiency.

shows the variation of the exergy yield of the pre-reformer as a function of the extraction fraction. This yield increases with the indicated fraction. For an extraction fraction equal to 0.05, the pre-reformer has an exergy yield equal to 23.9% while for a fraction fs equal to 0.1, this yield reaches 25.3%.

3. Effect of H_2 flow

The variation in the flow of H_2 with constant oxidant flow represents in other words the variation of the fuel/oxygen ratio. This ratio affects the electrochemical reaction, temperature, and combustion reaction.

At a given air flow and precisely at a given oxygen flow, corresponds a maximum flow rate of H_2 which could be oxidized without heating the cell to a temperature that damages the materials constituting the cell.

Figures 9 and **10** illustrate the influence of H_2 flow rate on the exergy efficiency of the fuel cell and gas turbine, respectively. It can be seen that for the SOFC, the efficiency increases significantly with the H_2 flow, reaching a limit value of about 53.4% as shown in **Figure 9**. For a variation in the H_2 flow rate from 101.1 to 203.3 mol/s, the exergy efficiency of the cell increases by approximately 8%.

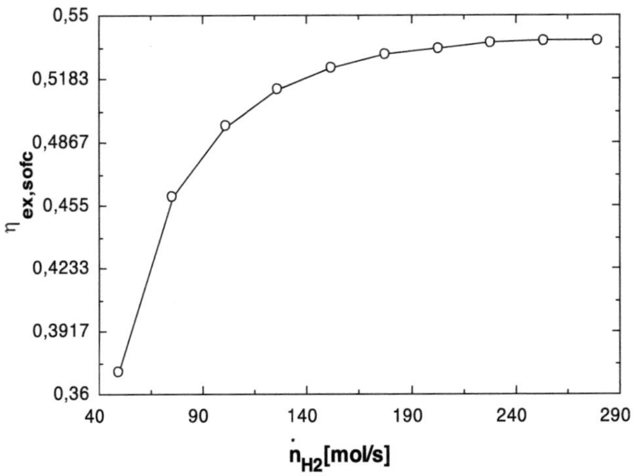

Figure 9.
Effect of H_2 flow on SOFC exergetic efficiency.

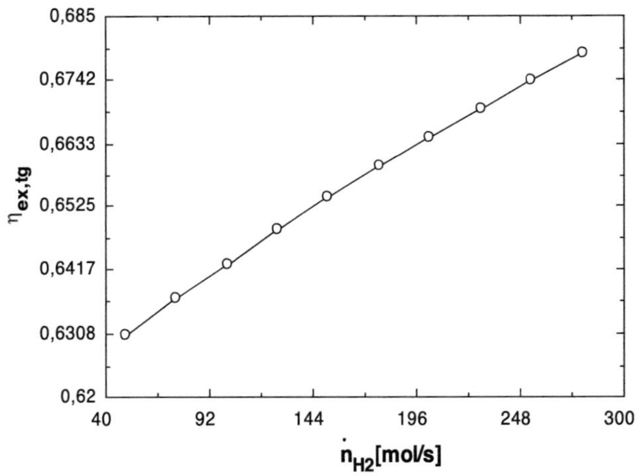

Figure 10.
Effect of H_2 flow on gas turbine exergetic efficiency.

Likewise, the exergy efficiency of the gas turbine is improved with increasing H_2 flow rate. For the same range of H_2 flow rate variation, the exergy efficiency of the gas turbine increases by approximately 3.2% as shown in **Figure 10**.

A positive effect was observed for the pre-reformer as shown in **Figure 11**. For the same variation in flow, the exergy efficiency obtained respectively are 14.4 and 24.4%.

We therefore find that irreversibility can be reduced by increasing fuel flow. This is due to the increasing of the cell temperature which decreases cell polarizations and to the increasing of the temperature and inlet flow of the gas turbine. We also note that the more the H_2 flow rate increases, the more the fuel flow rate increases and then more chemical exergy is available to the reformer.

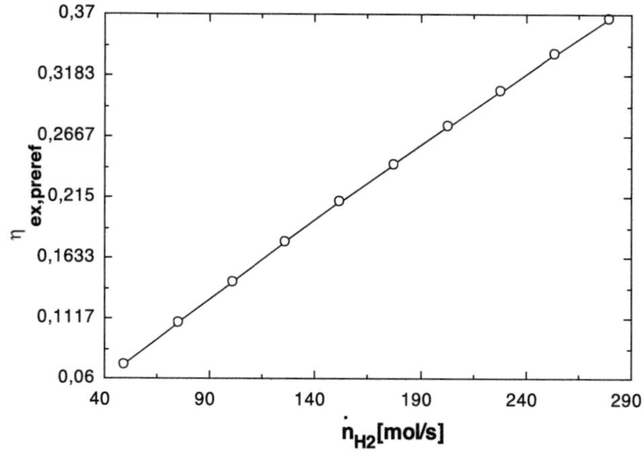

Figure 11.
Effect of H_2 flow on pre-reformer exergetic efficiency.

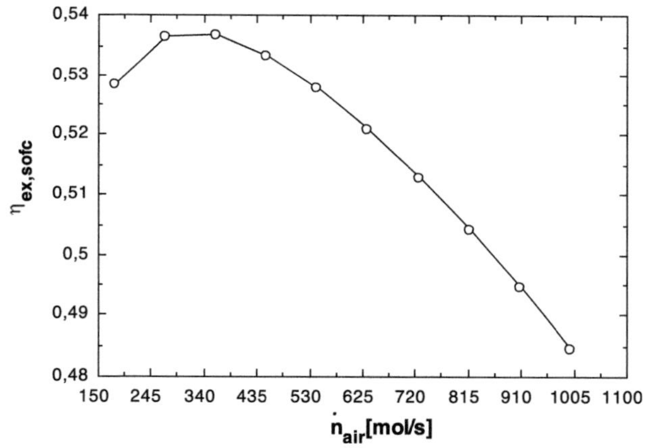

Figure 12.
Effect of air flow on SOFC exergetic efficiency.

4. Effect of air flow

The SOFC intake air flow affects cell, heat exchanger, and turbine performances. **Figure 12** shows the evolution of the SOFC exergetic efficiency as a function of the air flow. It is observed that the exergetic efficiency reaches a maximum value of approximately 53% for an optimum air flow rate equal to 362.2 mol/s. A higher air flow leads to a lower exergetic efficiency. This is explained by the fact that excess air cools the cell.

In addition, the more the molar air flow increases, the more the energy supplied to the cell increases, which leads to a higher exergy destruction.

In addition, the increase in the air flow leads to an improvement in the exergetic efficiency of the turbine as shown in **Figure 13**. For the same flow rate range, the gas turbine efficiency increases from 66.87 to 68.01% due to the increase in mass flow rate at the feed.

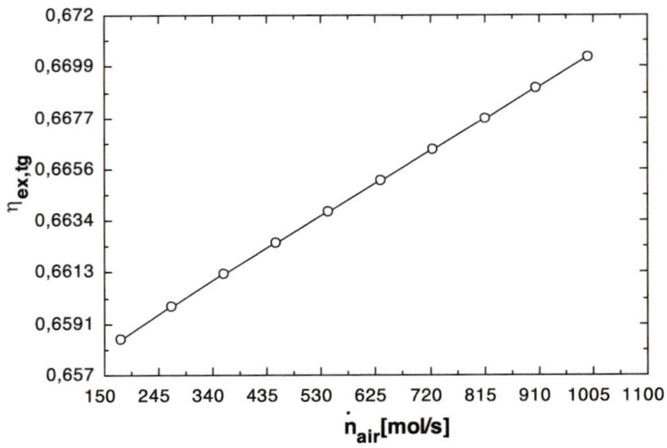

Figure 13.
Effect of air flow on the gas turbine exergetic efficiency.

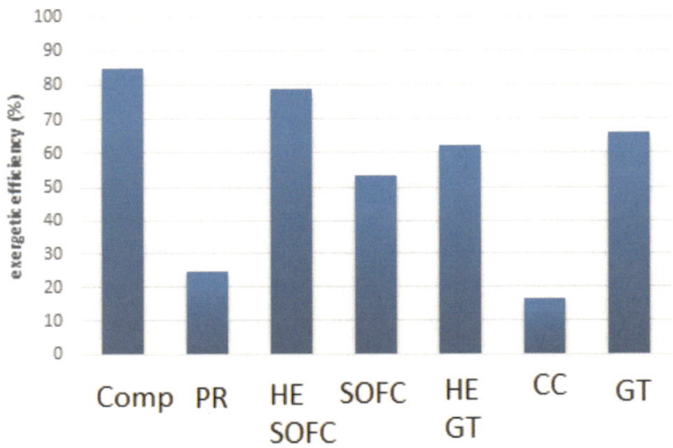

Figure 14.
Exergetic efficiencies of principal components.

5. Exergetic efficiencies of principal components in optimal conditions

The exergetic efficiencies of the various components operating under optimal conditions are shown in **Figure 14**. It is noted that the combustion chamber, the pre-reformer, and the fuel cell have the lowest values of the exergetic efficiency. Indeed, the combustion chamber has the lowest exergetic efficiency. It virtually destroys 83.3% of the exergy that is received. The irreversibility of combustion is due to heat transfer, friction of fluids, mixing, and chemical reaction. It is difficult to assess the contribution of each process to the total rate of irreversibility. In reality, we can assimilate that the irreversibility due to friction and mixing is negligible compared to other irreversibilities. The destruction of exergy at the SOFC is also significant. Its exergetic efficiency is in the order of 53.56% under optimal conditions. It is attributed to thermodynamic and electrochemical irreversibilities. Besides thermal transfer, SOFC is the location of various chemical reactions. It should also be noted that the reformer is a source of significant degradation of exergy since it is fed by material flows at different temperatures and it is the location of chemical reactions. While the destruction of exergy at the heat exchangers is explained by the difference in temperatures between the various circulating flows.

An exergetic analysis of the hybrid TG/SOFC system is developed. The exergy balances are established for the various components. The results obtained show that the exergy efficiencies are improved by increasing the pre-fixing fraction of the air and fuel flows. While the ambient temperature, humidity, fuel utilization factor, SOFC pressure, and the shrinkage fraction affect negatively the exergy performance of the hybrid plant. In addition, the exergy study made it possible to locate the inefficiencies of the system studied.

5. Conclusion

An exergy analysis of the hybrid TG/SOFC system is developed. The exergy balances are established for the different components. A parametric study made it possible to highlight the effects of the operating variables on the exergy efficiency of these components. The results obtained show that the exergy efficiencies are improved by increasing the pre-reforming fraction the air and fuel flows. While the pressure at the SOFC and the extraction fraction negatively affect the exergy performance of the hybrid plant. In addition, the exergy study made it possible to locate the inefficiencies of the system studied. The combustion chamber, pre-reformer, and SOFC have proven to be the greatest exergy destroyers.

Acknowledgements

The author is extremely grateful to the head of Applied Thermodynamic Research Unit, National Engineering School of Gabes, Tunisia, Prof. Tahar Khir for his encouragement and technical support.

Author details

Salha Faleh[1,2]* and Tahar Khir[2]

1 University of Hafr alBatin UHB, Saudi Arabia

2 Applied Thermodynamic Research Unit, National Engineering School of Gabes, Tunisia

*Address all correspondence to: sfaleh@uhb.edu.sa

References

[1] Sghaier SF et al. Energetic and exergetic parametric study of a SOFC-GT hybrid power plant. International Journal Hydrogen Energy. 2018;**43**(6):3542-3554

[2] Mahmoudi SMS, Khani L. Thermodynamic and exergoeconomic assessments of a new solid oxide fuel cellgas turbine cogeneration system. Energy Conversion and Management. 2016;**123**(1):324-337

[3] Valerie E, Wirinya K, Peter R, Ali AA. Energy, exergy and economic analysis of an integrated solid oxide fuel cell e gas turbine e organic Rankine power generation system. International Journal of Hydrogen Energy. 2016;**41**(31): 13843-13858

[4] Mahato N, Banerjee A, Gupta A, Omar S, Balani K. Progress in material selection for solid oxide fuel cell technology: a review. Progress in Materials Science. 2015;**72**:141-337

[5] Song TW, Sohn JL, Kim JH, Kim TS, Ro ST, Suzuki K. Performance analysis of a tubular solid oxide fuel cell/micro gas turbine hybrid power system based on a quasi-two dimensional model. Journal of Power Sources. 2005;**142**: 30-42

[6] Santhanam S, Schilt C, Turker B, Woudstra T, Aravind PV. Thermodynamic modeling and evaluation of high efficiency heat pipe integrated biomass GasifiereSolid Oxide Fuel CellseGas Turbine systems. Energy. 2016;**109**:751-764

[7] Park SK, Kim TS. Comparison between pressurized design and ambient pressure design of hybrid solid oxide fuel cell gas turbine systems. Journal of Power Sources. 2006;**158**:225-244

[8] Kuchonthara P, Bhattacharya S, Tsutsumi A. Combinations of solid oxide fuel cell and several enhanced gas turbine cycles. Journal of Power Sources. 2003;**124**(1):65-75

[9] Eveloy V, Rodgers P, Qiu L. Integration of an atmospheric solid oxide fuel cell -gas turbine system with reverse osmosis for distributed seawater desalination in a process facility. Energy Conversion and Management. 2016;**126**:944-959

[10] Yi JH, Choi JH, Kim TS. Comparative evaluation of viable options for combining a gas turbine and a solid oxide fuel cell for high performance. Applied Thermal Engineering. 2016;**100**:840-848

[11] Zhang X, Li J, Li G, Feng Z. Cycle analysis of an integrated solid oxide fuel cell and recuperative gas turbine with an air reheating system. Journal of Power Sources. 2007;**164**(2):750-760

[12] Lv X, Gu C, Liu X, Weng Y. Effect of gasified biomass fuel on load characteristics of an intermediate-temperature solid oxide fuel cell and gas turbine hybrid system. International Journal of Hydrogen Energy. 2016; **41**(22):9563e76

[13] Harun NF, Tucker D, Adams TA II. Impact of fuel composition transients on SOFC performance in gas turbine hybrid systems. Applied Energy. 2016;**164**: 446-461

[14] Yang L, Han J, You H. Performance analysis of a CCHP system based on SOFC/GT/CO_2 cycle and ORC with LNG cold energy utilization. International Journal of Hydrogen Energy. 2019;**44**(56):29700-29710

[15] Abdollahi HM, Ghazanfari HS, Ata C, Kiyan P. Thermodynamic assessment of a

novel multi-generation solid oxide fuel cell based system for production of electrical power, cooling, fresh water, and hydrogen. Energy Conversion and Management. 2019;**197**:111895

[16] Inac S, Unverdi SO, Adnan Midilli A. Parametric study on thermodynamic performance of a SOFC oriented hybrid energy system. International Journal of Hydrogen Energy. 2019;**44**:10043-10058

[17] Gholamian E, Habibollahzade A, Zare V. Development and multi-objective optimization of geothermal-based organic Rankine cycle integrated with thermoelectric generator and proton exchange membrane electrolyzer for power and hydrogen production. Energy Conversion and Management. 2018;**174**:112-125

[18] Malico I, Carvalhinho AP, Tenreiro J. Design of a trigeneration system using a high temperature fuel cell. International Journal of Energy Resources. 2009;**33**: 144-151

[19] Mehdi M, Hossein D, Ali MSM. Optimal design of solid oxide fuel cell, ammonia-water single effect absorption cycle and Rankine cycle. Journal of Power Sources. 2016;**306**:107-123

[20] Azizi MA, Brouwer J. Progress in solid oxide fuel cell-gas turbine hybrid power systems: System design and analysis, transient operation, controls and optimization. Applied Energy. 2018; **215**:237-289

[21] Pierobon L, Rokni M. Thermodynamic analysis of an integrated gasification solid oxide fuel cell plant with a Kalina cycle. International Journal of Green Energy. 2015;**12**:610-619

[22] Zhao P, Dai Y, Wang J. Performance assessment and optimization of a combined heat and power system based on compressed air energy storage system and humid air turbine cycle. Energy Conversion and Management. 2015;**103**: 562-572

[23] Sadeghi S, Ameri M. Exergy analysis of photovoltaic panels-coupled solid oxide fuel cell and gas turbine-electrolyzer hybrid system. Journal of Energy Resources Technology. 2014; **136**(3):031201 (10 pages)

[24] Valerie E, Wirinya K, Peter R, Ali AA. Energy, exergy and economic analysis of an integrated solid oxide fuel cell e gas turbine e-organic Rankine power generation system. International Journal of Hydrogen Energy. 2016;**156**: 32-85

[25] Meng Q, Han J, Kong L, Liu H, Zhang T, Yu Z. Thermodynamic analysis of combined power generation system based on SOFC/GT and trans critical carbon dioxide cycle. International Journal of Hydrogen Energy. 2017;**42**(7): 4673-4678

[26] Yang L, Jitian H, Huailiang Y. Performance analysis of CCHP system based on SOFC/GT/CO_2 cycle and ORC with LNG cold energy utilization. International Journal of Hydrogen Energy. 2019;**44**:29700-29710

[27] Xiaojing L, Lu C, Wang Y, Yiwu W. Effect of operating parameters on a hybrid system of intermediate temperature solid oxide fuel cell and gas turbine. Energy. 2015;**91**:10

[28] Yi Y, Rao AD, Brouwer J, Samuelsen GS. Analysis and optimization of a SOFC-ICGT intercooled GT hybrid cycle. Journal of Power Sources. 2004;**132**:77-85

[29] Haseli Y, Dincer I, Naterer GF. Thermodynamic analysis of a combined

gas turbine power system with a solid oxide fuel cell trough exergy. Thermochim Acta. 2008;**480**:1-9

[30] Harvey SP, Richter HJ. Gas turbine cycles with solid oxide fuel cells. Part I: improved gas turbine power plant efficiency by use of recycled exhaust gases and fuel cell technology. Journal of Energy Resources Technology. 1994; **116**(4):305-311

[31] Zhang X, Li G, Li J, Feng Z. Numerical study on electric characteristics of solid oxide fuel cells. Energy Conversion and Management. 2007;**48**:977-989